Farming and the Burren

Brendan Dunford

Teagasc
Agriculture and Food Development Authority
www.teagasc.ie

Foreword

In terms of its natural and cultural heritage, the Burren landscape is a veritable oasis. The Burren uplands play host to over 70% of Ireland's native flora, hundreds of burial sites, stone forts and ancient ecclesiastical sites. It is a key flagship for the Irish environment, remarkably rich, diverse and intact.

However, the Burren is a landscape that depends very much on farming activity for the retention of its rich heritage. For over six thousand years, farmers have shaped the Burren. In much the same way, the unique challenges presented by karst landscape of the Burren have shaped use of the land and evoked distinctive agricultural responses.

Many of these land management traditions are now at risk, as the balance that evolved between farming activity and the environment of the Burren over many millennia is being threatened by the rate and scale of recent changes in the agricultural sector. Sustainable solutions to redress this balance must now be found, and these solutions must be acceptable to both farmers and conservationists.

Teagasc is deeply committed to the sustainable future management of the Burren, and to the future of the several hundred farm families who depend on it for a living. Brendan Dunford, the author of this book has participated in a four year Teagasc sponsored post-graduate research study. This is the outcome of his work. I congratulate him on a superb product and am confident you will find it entertaining and thought provoking.

Jim Flanagan, Director, Teagasc.

Brendan Dunford has spent the past five years in the Burren region investigating the evolution and significance of the relationship between the rich heritage of the Burren and the farming communities of the region. He was conferred with a PhD, based on these studies, in 2002. He continues to reside in the Burren and currently works with Teagasc in Clare looking at ways of securing a sustainable future for the farmers of the Burren and their prized heritage.

Published by Teagasc, December 2002
Designed and typeset by DBA Publications Ltd, 56 Carysfort Avenue, Blackrock, Co. Dublin
Printed by Grafton Litho Ltd.

Farming and the Burren

-the story of farming's contribution to one of Europe's most precious landscapes

Contents

1.	Introduction	1
2.	The Burren: an overview	4
3.	Burren farming through the ages	21
4.	Traditional Burren farming practices	32
5.	Burren farming today	51
6.	The ecological impact of agricultural practices	72
7.	The future: threats and promise	89
8.	Acknowledgements	101

Brendan Dunford

December, 2002

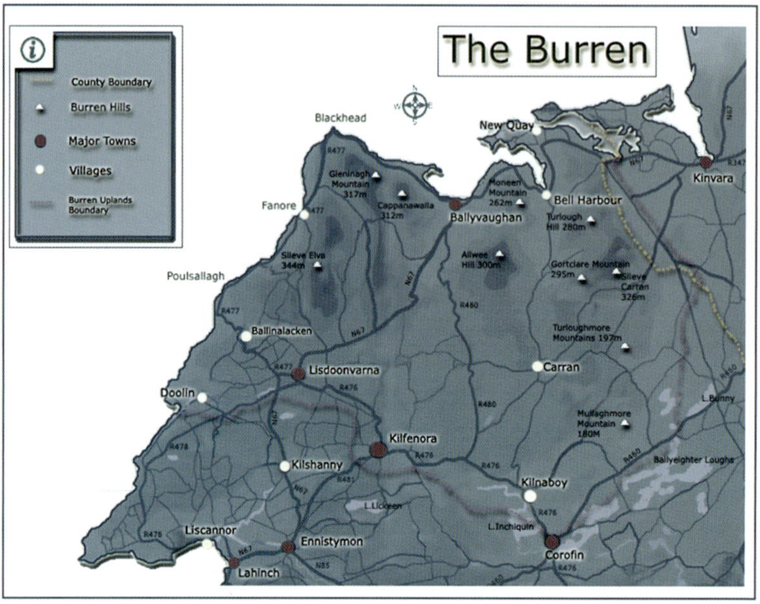
The Burren
County Boundary
Burren Hills
Major Towns
Villages
Burren Uplands Boundary
Blackhead
New Quay
Kinvara
Gleninagh Mountain 317m
Cappanawalla 312m
Fanore
Ballyvaughan
Moneen Mountain 262m
Bell Harbour
Turlough Hill 280m
Slieve Elva 344m
Aillwee Hill 300m
Gortclare Mountain 295m
Slieve Carran 326m
Poulsallagh
Ballinalacken
Turloughmore Mountains 197m
Lisdoonvarna
Carran
Doolin
L.Bunny
Mullaghmore Mountain 180M
Kilfenora
Kilshanny
Ballyeighter Loughs
Kilnaboy
L.Lickeen
L.Inchiquin
Liscannor
Ennistymon
Corofin
Lahinch

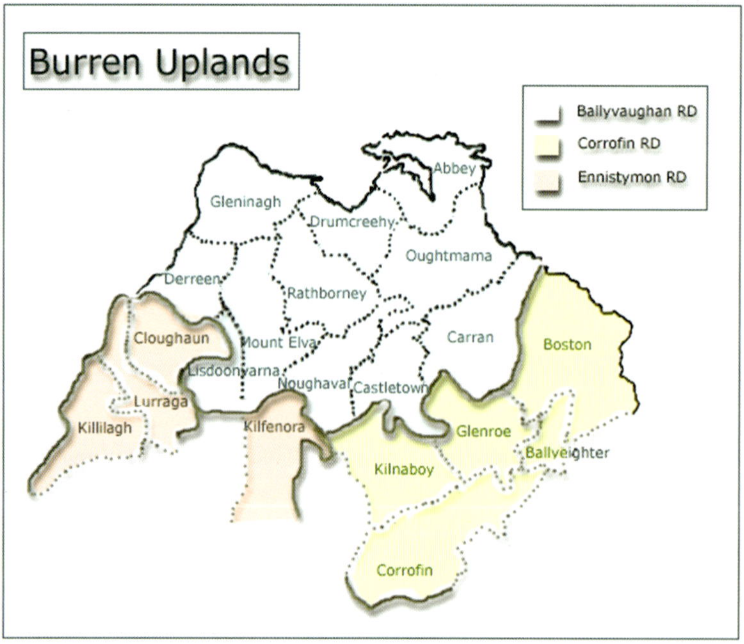
Burren Uplands
Ballyvaughan RD
Corrofin RD
Ennistymon RD
Abbey
Gleninagh
Drumcreehy
Oughtmama
Derreen
Rathborney
Cloughaun
Mount Elva
Carran
Boston
Lisdoonvarna
Noughaval
Castletown
Lurraga
Killilagh
Kilfenora
Glenroe
Ballyeighter
Kilnaboy
Corrofin

1. Introduction

The Burren upland region is, without doubt, one of Europe's most important, distinctive, widely recognised and best loved landscapes. This international reputation is somewhat at odds with the harsh, barren appearance presented by the limestone hills that characterise the region, and their remote location on the north-western edge of the European seaboard. However the reputation of the Burren has, if anything, grown of late, the unique natural and cultural attributes of its enigmatic landscape increasingly valued in a Europe where much of the countryside has lost its intrinsic character, diversity and appeal. More and more people are drawn, physically and emotionally, to this resilient, uncompromising place to rekindle their relationship with the land, with nature, with their heritage.

The distinctive landscape of the Burren uplands, which extend over c. $360km^2$.

The Burren is special by any number of standards. Notwithstanding its dramatic beauty, it is, undoubtedly, one of Europe's finest examples of what is described as a 'glaciated karst' landscape. These hills are home to a remarkably diverse and well-preserved built heritage that is richly representative of every era from the Neolithic period to the present day. Most famously perhaps, the Burren also boasts a stunning flora, the sheer beauty, diversity and unique attributes of which have attracted legions of admirers.

While each of these features has been subject to much well-deserved attention and investigation, one very special and integral aspect of the Burren's multifaceted landscape has somehow escaped such perusal: the fascinating story of farming in this region and its contribution to every aspect of this rich and varied heritage. Our understanding and appreciation of the Burren is incomplete without a greater awareness and appreciation of this remarkable, and quite intriguing, story.

Innumerable generations of farmers have left their mark on the Burren, influencing the area as they struggled to wrest a living from this challenging landscape, the particular attributes of which often forced farmers to adopt somewhat unusual management practices, many of which still survive. This human influence pervades the Burren upland landscape, reflected in the barren hillsides criss-crossed with bare stone walls, and in the innumerable monuments, be they funerary or functional, that are liberally strewn throughout. The impact of farming is also reflected in the phenomenally rich flora of the upland grasslands, the presence of which is closely related to the ancient tradition of transhumance, known locally as 'winterage'. Today's generation of Burren farmers continue to uphold the traditions of their forebears, though influenced to an unprecedented degree by external political, market and technological forces

The regal flowering heads of a Burren favourite, the golden Carline thistle.

In 1998, Teagasc initiated a research programme to investigate the impact of agricultural practices on the natural heritage of the Burren uplands, a reflection of the growing concern for this rich heritage and its custodians, and the pressures that recent changes have brought on both. Some of the findings from this research programme are presented in this book. It is hoped that this will serve to increase our understanding and promote broader acknowledgement of the immense contribution made by farmers to the heritage of the Burren. This will, hopefully, encourage a greater sense of understanding, pride and responsibility towards this special place among all those involved in its management, and contribute towards securing the future of what is, in a modern context, an increasingly precious and threatened resource.

Grass of Parnassus

2. The Burren: an overview

Geology

The characteristic limestone hills that earned the Burren its name (from the gaelic word *Boireann*, or rocky place), were laid down some 360 million years ago, during what is known as the Carboniferous period. This visibly fossil-rich limestone is composed of the compacted ancient remains of the life – shellfish, corals, and many other microscopic organisms - of the warm shallow seas that once covered this region, at a time when it lay south of the equator. Up to 800m thick in places, this limestone is arranged in distinctive tiers or bands that are separated by lines of weakness in the limestone, or in some cases by thin layers of shale. These impermeable shale layers are the remnants of muds that periodically washed into the ancient seas from adjacent land masses. A thick cover of this Namurian shale was eventually deposited on top of the Burren limestone, and though glaciation stripped much of it away, outcrops still remain at Slieve Elva and Poulacapple, contrasting starkly with the surrounding limestone (see Simms, 2001).

Karren features called rinnenkarren, or solution runnels.

Through the action of glaciers, and the solutional effects of rainfall, the Burren's limestone has been, and continues to be, sculpted over time into a rich array of unusual forms. Among the most characteristic features are rounded hollows of varying size called *dolines*, disappearing lakes called *turloughs*, and innumerable networks of caves. These are collectively referred to as *karst* features. At a more minute scale, the mild natural acidity of rainfall gradually burns through the limestone to form innumerable etchings – ridges, hollows and runnels – collectively referred to as *karren* features. The Burren is so rich in these glacial, karst and karren features that it is widely recognised as one of the finest examples of a 'glaciated karst' landscape in Europe. Even in the absence of its rich flora and archaeology, the geology of the Burren would in itself merit special distinction, and is a major source of attraction for geologists and university students from all over the world.

A glacially deposited limestone boulder or *erratic* in Fanore. Note the pedestal underneath, indicating the extent to which the surrounding pavement (made up of massive blocks called *clints*, separated by fissures or cracks called *grikes*) has been weathered away through several Millennia of rainfall dissolution.

It is important to understand that many of these highly individual and unique geological features have been shaped over tens of thousands of years by natural forces. Their disturbance or removal for building or reclamation is thus very damaging, particularly so when one considers how limited this resource is elsewhere in Europe. For instance, only 2900ha of limestone pavement remain in Britain, and 97% of this has already been damaged according to a recent report (Traffic International, 1999).

While in the Burren limestone rocks would historically have been mined by hand to build tombs, forts, walls and roadways, the scale and severity of rock removal by mechanical excavation is potentially much more extensive and damaging. With a few lamentable exceptions, the state-subsidised agricultural 'reclamation' of areas of limestone pavements that occurred in the 1980s (Drew and Magee, 1994) has ceased. Sadly, the removal of limestone pavement continues however, driven primarily by the upsurge in demand for the use of 'water worn' limestone in the construction and horticultural industries. Elsewhere, the construction of mini-dolmens, wanton smashing and removal of limestone rocks, and even the toppling of tenuously perched glacial erratics by misguided visitors is also a damning indictment of that industry. Such interference with the very precious geological heritage of the Burren is currently outlawed in NHA/SAC areas, which encompass much of the region (Dúchas, 1999).

Limestone pavement, Cappanawalla.

Soils

It is now generally believed that, prior to man's arrival in the Burren, a rich mineral soil cover existed which supported an extensive woodland of pine, with an understory of hazel. Attracted by these light fertile soils, industrious Neolithic and Bronze Age farmers gradually cleared the woodland for cultivation. However, without the tree roots to bind it in place, and subjected to quite an intensive degree of agricultural use (according to archaeological evidence), it appears that much of this soil was washed away when the climate deteriorated, sometime in the late Bronze Age, roughly 3,000 years ago. Thus, the characteristic barren hills that later merited the title *Boireann* can in fact be viewed as a stark reminder of man's fundamental contribution to the evolution of this landscape, and of the vulnerability of the soil resource that we sometimes take for granted.

Typical Burren species such as blue moor grass, early purple orchids and bird's-foot-trefoil, prosper on the thin, rocky rendzina soils found in the Burren uplands.

The soils that sporadically clothe the skeletal limestone of the Burren today are widely varied in type and quality. The main soil groups in the Burren were mapped as part of the county survey by Finch (1971). The most common soil type is described as a rendzina: a very thin (usually <5cm), dark, free-draining organic soil that dominates many upland grasslands. Rendzinas are quite limited agriculturally, but are well suited to winter grazing, as they are very resistant to poaching, and remain relatively dry and warm throughout the year. These thin soils retain little moisture, are usually heavily leached, and thus nutrient-poor, facts of enormous ecological significance. Conventional grassland species lose their competitive advantage on such limiting soils, leaving the way open for stress-tolerant species, many of them Burren specialities, to prosper..

Elsewhere in the Burren, deeper and more fertile mineral soils are found, usually deposited by glaciers (running north-east to south-west) along valleys or hillsides, often forming distinctive green islands or pockets that contrast with the surrounding rocky grey expanse. These boulder clay soils are quite free draining and fertile, eminently suitable for tillage or silage production or year round grazing. The distribution and extent of other mineral soils, such as wind-blown loess, remains subject to much question and debate. Though these mineral soils are usually the focus for quite intensive farming activity, patches that have escaped improvement often support a very interesting flora.

Occasionally, small pockets of deeper soils are found scattered high in the rocky uplands. Many such pockets would have been walled off and cultivated by hand, these so called *gairdini* (little gardens) producing crops of potatoes, oats or hay, critical to the livelihood of the small farmer or herdsman. In some upland areas where drainage is obstructed, heathy mounds of soil and vegetation develop: in many cases these patches would have been harvested for fuel, the sods dried out by placing them in parallel stone wall structures called *thuiles* through which the wind could blow freely. On occasion, *thuiles* would have been used for drying out cow-dung pats, also used for fuel.

Hundreds of these, and many other, mysterious walled structures, can be still found high in the Burren hills, testament to the ingenious and detailed exploitation of the local soil resource. They constitute a fitting symbol of the struggle for existence waged by the farmers of these hills in the past, and also of the rich, extensive, though largely unrecognised, legacy of these lost generations.

Thuiles and a shelter wall (centre left) high in the Burren, symbols of a harsh past, and of the largely forgotten agricultural heritage still to be found in the uplands.

Hydrology

It has been claimed, with some justification, that water, not rock, is the essence of the Burren. This is ironic in light of the apparent lack of water in the region, with only one consistent overground river present, the Caher in Fanore. Most of the other water courses in the Burren have worked their way underground over thousands of years, forming labyrinthine networks of caves, both fossil and active. Some of these water flows reappear suddenly and briefly after heavy rainfall. Others, such as turloughs (from the gaelic words *tuar loch* meaning disappearing lake), are usually slower to respond, rising and falling in accordance with fluctuations in the water table, often leading to their use locally as a crude barometer of recent rainfall levels (see Drew, 1990).

A periodically submerged stone wall crossing a turlough, following several weeks without rain.

The influence of water is everywhere to be seen in the Burren. A combination of high rainfall levels (c.1500mm annually) and the high solubility of the limestone has contributed, over time, to the wealth of distinctive karst and karren features that exist in the region. The drought-like conditions of the uplands in summer have also evoked a distinctive ecological response – the number of species of Mediterranean origin found prospering in the region for example - while turlough areas also contain many interesting species that have adapted specifically to their unusual environment.

Agriculturally, the stocking restrictions caused by the limited water supply in the uplands must have contributed to the evolution of the Burren's winterage tradition, and the attendant ecological and cultural implications. Farmers traditionally responded to the inconsistent water supply in the Burren by building elaborate rainwater-collection tanks, still visible in upland areas to this day. Today however, these limitations are more easily circumvented by piping water to the uplands, contributing in some cases to a change in management of these areas, and thus their flora.

Rainwater collection and storage tank. The size, number and distribution of these tanks in the Burren uplands reflects the scarcity, and value, of a consistent water supply, a situation typical of karst landscapes all over the world.

The unusual hydrology of the Burren does however pose certain challenges for today's farmer and rural dweller. Because of the thin soils in the Burren, and the speed at which water passes through the limestone bedrock, there is very little attenuation or trapping of water-borne pollutants. This is why such extreme care is warranted in the Burren in terms of sewage disposal, fertiliser and slurry application, feeding site and fodder management. The high degree of susceptibility of the all-important groundwater resource to pollution in the Burren explains why nutrient management planning and pollution control measures are among the most significant elements of the Burren Rural Environmental Protection Scheme (REPS).

Lough Bunny, east Burren.

Flora

The flora of the Burren is probably its most renowned feature, attracting thousands of visitors annually to delve in its extraordinary richness. Some 635 (70.5%) of Irelands 900 native plant species have been found in the Burren hills (Webb and Scannell, 1983), though these hills represent less than 0.5% of the national area. Included among them are 23 of Ireland's 27 native orchid species, and distinctive specialities such as the widely recognised spring gentian, mountain avens, harebell, ladies' bedstraw and bloody crane's-bill, to name but a few. Many of these plants have their main British Isles populations in the Burren, while several other species 'flourish so much more exuberantly in the Burren than elsewhere in Ireland' (Webb and Scannell, 1983).

It is no coincidence that almost all of the more interesting species and communities of Burren flora are found on winter grazed upland grasslands. The thin, nutrient-poor, unimproved soils and winter grazing regime serve to prevent the more competitive plant species and litter from monopolising the vegetation, while the absence of animals in the flowering season allows the often-minute herb flora to prosper unhindered. Animal's hooves also create patches for seedling establishment and grazers redistribute nutrients and seedlings through dunging, further enhancing floral diversity.

Bee orchid, found widely in the Burren uplands. This fascinating plant (closely resembling a species of bee in structure, texture and colour) is but one of a stunning collection of orchid species found in the region.

The seasonal transformation of the ostensibly drab, barren uplands in winter into a multicoloured summer tapestry is nothing short of miraculous, inspiring the Cistercian monks at Corcomroe to dedicate their Abbey to 'Our Lady of the fertile rock', representing, in the words of MacMahon (1997), 'an exultation of the environmental and ecological paradox that is the Burren'. Without grazing animals, there is strong evidence (section 6) to suggest that the Burren's rich plant diversity is severely diminished, as rank grasses and eventually scrub take over. The Burren flora as we know it can not, realistically, survive in the absence of farmers. Ironically, as more farmers join REPS, and others diversify into tourism, there is also a growing realisation among many farmers that their livelihood, their survival, is increasingly dependent on working to protect this very heritage.

Typical species-rich Burren grassland featuring mountain avens (white seed heads), bird's-foot-trefoil (yellow) and bloody crane's-bill (purple).

To promote and support this role of farmers as custodians of the countryside is one of the cornerstones of the new concept of multifunctional agriculture envisaged by the EU for its farmers. Burren farmers, through the wealth of their heritage and tradition, are in the front line of this new crusade. It is not unrealistic to anticipate that, in future, the 'environmental goods' produced by these farmers – be they species rich grasslands or well preserved archaeological sites – will surpass the 'agricultural goods' produced, in terms of their significance and value to the overall farm economy.

Fauna

Though frequently overlooked, the fauna of the Burren uplands is also very rich and varied. The most visible mammals found in the uplands are the fox (*Vulpes vulpes*), hare (*Lepus timidus*) and feral goat (*Capra hircus*). Woodmice (*Apodemus sylvaticus*), red squirrels (*Sciurus vulgaris*) and badgers (*Meles meles*) are also very common, though more difficult to spot. All of Ireland's seven native bat species are found in the Burren, including a very significant population of the lesser horseshoe bat (*Rhinolophus hipposiderus*). Over seventy species of land snails have been recorded in the area, thanks in no small part to the abundance of shell-building calcium carbonate.

The Burren has more than it's fair share of unusual species, including the pine marten (*Martes martes*), slow worm (*Anguis fragilis*), green lizard (*Lacera viridis*) and Burren green moth (*Calamia trideus*), all of which benefit from the unique habitats available in the region. All but two of Ireland's 30 or so butterfly species are found in the Burren, with two more or less limited to this area - the pearl bordered fritillary (*Boloria euphrosyne*) and the brown hairstreak (*Therla betulae*). A significant bird population is also found, including several pairs of nesting peregrine falcons (*Falco peregrinus)* and the large number of wintering wildfowl found in turlough areas (see D'Arcy and Hayward, 1992)

Transparent burnet moth on bloody crane's-bill flower.

Some of these species are viewed as a threat by farmers. Pine martens and foxes may kill lambs. Badgers are frequently blamed for the transmission of bovine TB. The feral goat population, numbered in thousands, is among the highest in Europe, and represents a problem for many farmers, as they knock walls (exposing the REPS farmer to a 20% penalty) and reduce available grazing for livestock through spoiling and browsing. This is a useful example of how farming and conservation interests may sometimes clash, a conflict that can and must be accommodated and addressed by the relevant agri-environment policies and schemes.

Feral goat herd in the north-east Burren.

Built heritage

The Burren, though barren by name and appearance, is paradoxically rich in built heritage. Cartographer Tim Robinson described the Burren as one 'vast memorial to bygone cultures' (Robinson, 1999), such is the wealth and diversity of heritage in the region. Monuments from every era since the Neolithic grace the Burren hills, testament to the attraction that this area has always held for farmers, and the ample supply of building stone readily available. The inherent unsuitability of these hills to the plough or bulldozer helped to preserve these monuments, many of which were sacrificed elsewhere in Ireland in the process of reclamation and cultivation.

The richness and diversity of the Burren's built heritage is truly stunning. Some 75 of Ireland's 400 wedge tombs are found in the Burren, as are c. 500 ring forts and several hundred *fulacht fiadh* (Bronze Age cooking sites). The Burren is said to contain Ireland's highest density of ecclesiastical sites, with spectacular examples at Oughtmama, Corcomroe and Kilfenora. One of Ireland's most recognisable monuments, Poulnabrone dolmen, is found in the heart of the Burren, a few miles north-west of one of its greatest hill forts, the triumvallate (three-walled) Cahercummaun.

Wedge tomb, central Burren.

Of lower profile, but no less interesting, are the vast assortment of traditional farming structures that grace the Burren uplands, contributing to what must be regarded as one of the most representative and well-preserved agricultural landscapes in Europe. *Cahers* and *crós* for rearing goat kids, turf *thuiles*, sheep folds and *puickets*, herdsmen's stone cottages, rainwater collection tanks, famine roads and winding, hand-cleared pathways all survive in profusion, reflecting the extent to which, and ingenuity with which, these hills were once farmed. Rather than detract from the natural beauty of the Burren, many of these man-made features embellish it, adding a fascinating new dimension to an already extraordinary landscape.

The fact that the Burren hills boast such a diverse, representative wealth of monuments, many of which appear to blend seamlessly into the surrounding landscape, has sometimes contributed to their destruction and neglect. Many of these structures have retreated from the mental landscape also, as countless anecdotes relating to their origin and use have been lost forever, the importance of these fascinating little histories scarcely appreciated as the treasure that they actually represent. All of us, farmers, visitors and policymakers, must work to recognise, understand and cherish this heritage and its associated folklore, as befits such a precious and rare resource.

Examples of some traditional farming structures in the Burren. On the left is a sheep *puicket*, a simple construction used to allow sheep access between grazing areas while denying cattle such movement. On the right, one of the Burren green roads, once used primarily to facilitate the transfer of livestock, but now more popular for recreation and leisure.

Agriculture

Belying its barren appearance, and much to the surprise of the majority of visitors to the region, the Burren sports a very rich agricultural tradition that extends over several thousand years. Several hundred farm families continue to uphold this tradition today, though the systems they employ have undergone radical change, particularly over the past few decades. Propelled by irrepressible international market, political and technological forces, these changes have upset the balance that traditionally existed between farming and the local environment in the Burren, and are increasingly threatening what was for many generations, a largely symbiotic, and very productive (ecologically, culturally and agriculturally) relationship.

In recent years Burren farming has become more specialised, intensive and concentrated in the fertile valley regions that intersect the grey limestone hills. Though the agricultural sector remains the biggest employer in the region[1], the numbers employed in the sector have dropped by 52% over the period between 1970 and 1996.

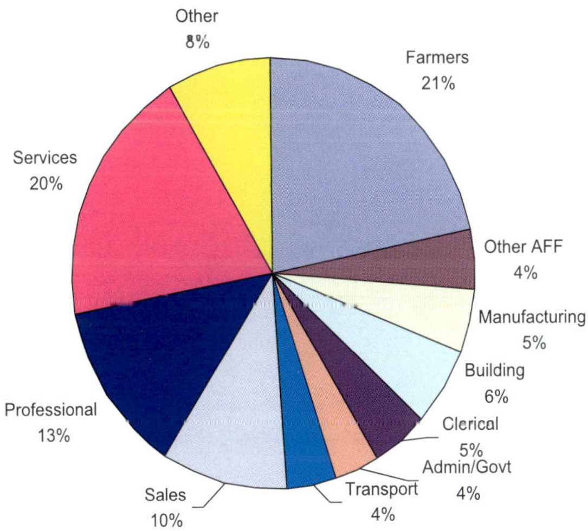

Employment by industry sector 1996, Ballyvaughan RD (source: CSO, 1998). The importance of agriculture to the local economy is apparent: some 25% of jobs are directly related to Agriculture, Forestry, and Fishing (AFF). The next most important sector is services, often associated with tourism, and particularly important in the towns of Lisdoonvarna and Ballyvaughan.

[1] The Burren region, though lacking accurate political definition, is generally considered to encompass roughly 20 District Electoral Divisions (DEDs), or parts thereof, within three Rural Districts (RDs). Of the three RDs – Ennistymon, Corofin and Ballyvaughan -the latter is most closely associated with the Burren, encompassing 75% of its area. For the purposes of this study, data relating to Ballyvaughan RD is used for statistical comparisons, to reflect the situation in the broader Burren region.

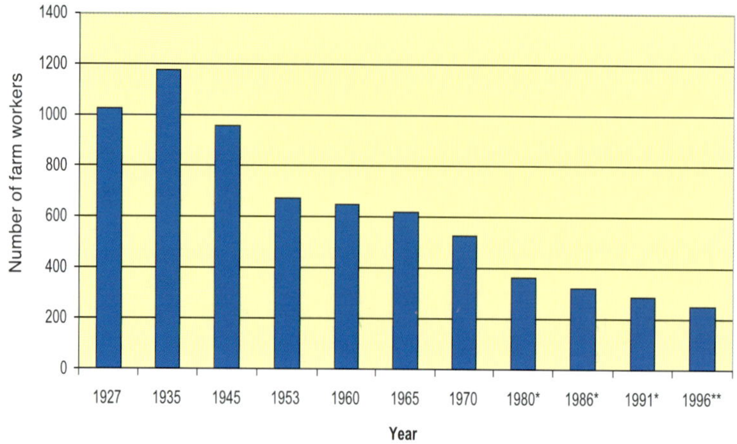

Number of males employed in agriculture in Ballyvaughan RD, 1927-1996 (source: CSO assorted dates). (Note: Both *[Brady Shipman Martin, 1996] and ** [CSO, 1998] refer to number of males *and* females employed in the AFF sector).

This erosion of the labour supply reflects the diminishing viability of the agricultural sector, the lure of the buoyant off-farm economy and the consolidation of land holdings. Increasing levels of mechanisation have resulted, and many of the less accessible upland areas of the Burren have tended to become marginalised in the increasingly streamlined farm economy, with ancient traditions such as 'winterage' becoming corrupted.

Elsewhere, the process of land marginalisation and abandonment has contributed to the afforestation of some of Europe's finest karst landscapes, including the proverbial Slovenian Kras region, from where the term karst originates. This suggests that a discontinuation of agriculture and its associated traditions will greatly devalue the conservation importance of the Burren, unless solutions are found. This is a conclusion strongly supported by the findings of this study.

The Burren farming community are justifiably proud of their heritage and traditions. They know their patch of the Burren with an intimacy and depth that many self-professed experts can only aspire to, and they alone have the unique capacity to evoke the rich human histories hidden in the stones. However, with this knowledge comes a great responsibility - to continue the guardianship of former generations in order that future generations can also have the opportunity to appreciate this place and earn a living from it. Through appropriate policies, research and education, and with the understanding and support of the broader community, this can, and indeed must, be achieved.

3. Burren farming through the ages

Neolithic farmers

Evidence from archaeological excavations and pollen studies suggest that as far back as 6,000 years ago, farmers were actively impacting upon the heritage of the Burren uplands. These early Neolithic farmers would have had a mixed stock of cattle, sheep, goats and pigs, and would have supplemented their subsistence economy with some cereal cultivation, hunting and gathering. Their farming system would have been characterised by semi-nomadic phases of woodland clearance, cultivation and abandonment, concentrated in the relatively easily worked soils of the uplands. The light pine-hazel woodland that would have covered the uplands at this time would have been a considerable attraction for these farmers, due to its ease of clearance, with polished stone axes, and possibly fire, employed for this purpose.

Generations of these early farmers were interred in the portal, court and wedge tombs that remain visible today, particularly in the south of the Burren. According to de Valera and Ó Nualláin (1961) 'the dominant factor in the economy of the tomb builders in Clare was pasturage ... cattle would appear to have been the principal stock, though sheep and goats could also have played an important role'. The excavation of Poulnabrone portal tomb, built 5,800 years ago, revealed evidence of 22 internments and of a mixed farm economy based on cattle, sheep, goat and cereal (Lynch, 1988).

Poulnabrone Portal Tomb, 5,800 years old.

Mound walls. Extensive networks of these ancient structures are still to be found in the Burren, particularly in the southern uplands.

As time progressed, these farmers became more stable, organised and prosperous, constructing walled settlements, the remains of which are quite common, though easily overlooked, appearing only as small mounds snaking across the landscape. One such settlement, at Roughan hill, south of Carran, spanned 175 acres, encircled by these 'mound walls' and centred on a circular house and yard. Recent research on this and other sites has confirmed that the Burren upland region was 'a place densely settled and intensively farmed' some 4,500 years ago, at the end of the Neolithic period (Jones, 1997).

Farming in the Bronze Age – Iron Age

This prosperity continued into the Bronze Age (c.2500-500BC) that followed. Hundreds of 'cist' or 'box' graves were built during this period, many covered with enormous mounds of rock. The excavation of Poulawack cairn near Kilcorney in 1935 revealed ten separate cist graves and the remains of sixteen people, while hundreds of other cairns such as those on Turlough hill and Slieve Carran remain unexplored.

Poulawack multiple cist cairn.

Several hundred *fulachta fiadh* found in the Burren also date from around this time. Always found close to a water supply, these inconspicuous horseshoe-shaped mounds are in fact ancient cooking (and possible bathing) sites. The mounds are composed of discarded burnt limestone used to transfer heat from a nearby fire to the water in a central cooking pit. At the excavation of such a site near Carran, a sheep was cooked (and consumed) very efficiently using this ancient method (Ó Drisceoil, 1988). Cattle, deer and horse bones and an oak trough were also recovered at the Carran site, estimated to originate from over 3,000 years ago, testament to the agricultural traditions of the time.

Fulacht fiadh, Slieve Carran Nature Reserve. Note the presence of a (dry) water source in the foreground.

Pollen studies confirm that there was a 'drastic increase in the number of clearings' in the Burren during the Bronze Age (Feighan, 1985). The use of fire may have been a factor – hazel charcoal fragments dated to the late Bronze Age have been recovered within the Burren. It is now believed that these clearances, most likely in conjunction with a subsequent deterioration in the climate, resulted in the extensive loss of soil in the Burren uplands, forming the spectacular semi-natural landscape we see today.

During the Iron Age that followed (500BC-500AD), farming activity appears to have been greatly diminished. The limited extent of the archaeological heritage available from this period – some ring barrows, forts and fulachta fiadh – would support the notion of an 'Iron Age lull' in farming activity. Pollen studies reveal an extensive regeneration of secondary hazel woodland in the Burren at this time, an interesting prehistoric precedent for the current trend of hazel scrub regeneration in the region.

Early Christian Period farming

At the beginning of the Early Christian Period (c. 500 AD) agriculture began to recover strongly from the lull it had experienced over the previous millennium of the Iron Age. Written records from this period reveal that tributes composed of beef cattle, cows, sheep, pigs and cloaks were paid to the kings of Ireland from the tribes of the Burren. Agriculture at this time would have been boosted by the introduction of new technologies, particularly those relating to crop husbandry and milk processing. New crop species, tools and cultivation techniques were probably introduced to the Burren by monastic settlements such as those at Oughtmama and Corcomroe.

Significant advances in dairying were also taking place. Heretofore cows would have been used primarily for breeding and as a meat source, with many cattle butchered prior to the winter to provide food when the milk supply had dried up. However, with the advent of milk processing in the form of cheese and butter, cows could provide year round sustenance, a fact reflected in the perception of cows as 'the new form of capital that produced interest in the form of milk, cheese and butter' (McCormick, 1995). To protect these highly coveted animals from marauding wolves and raiding parties, ring forts or *cahers* were built - over 500 in the Burren alone. These Early Christian farmsteads again noticeably favour the Burren uplands in terms of location, though a number of earthen forts or 'raths' are also found in adjacent lowland areas.

Stone fort or Caher, west Burren, one of the many ancient farmsteads littered across the Burren uplands, testament to the high historic levels of farming activity in the area.

At Cahercommaun ring fort – one of Ireland's great stone forts - dated to c.800AD by a team from Harvard University in 1934 (Hencken, 1938), extensive evidence confirming the prevalence of a mixed farm economy was recovered. Bones of cattle, sheep, goats and red deer were identified, along with grinding (quern) stones for grain. The large amount of spindle whorls that were also recovered suggests that the fort may even have served as a major wool-processing centre for the region (see Cotter, 1999). This evidence again attests to the historical importance of dairying and sheep husbandry in a land often misguidedly associated exclusively with the winter grazing of store cattle.

It is interesting to note that many early monastic sites in the Burren, such as Oughtmama (c.1000AD) and Corcomroe (c. 1200AD), were strategically located between upland and lowland areas, reflecting the dual requirements of the mixed farm economy – grazing and tillage – of this time. It was only during this Early Christian period that lowland areas in the Burren began to be seriously exploited, most likely to meet the needs of tillage production that the uplands could not provide. From this period onward, the centre of activity appears to have gradually shifted to these fertile lowlands. It is noteworthy that ecclesiastical sites, of which the Burren is said to contain the densest concentration in the country, are predominantly located on areas with the richest soils, in contrast with the upland bias of previous constructions, (Mytum, 1982, Ní Ghabhláin, 1995).

Corcomroe Abbey, situated in a fertile valley at the foot of Abbey hill in the north Burren. Fittingly, the monastery was dedicated to 'Our lady of the fertile rock'.

Medieval times

That the richness of the Burren in terms of its agricultural produce was of historic renown is attested to by the frequency of the raids to which the region was subjected in medieval times. Tales of daring raids to the Burren feature frequently in the pages of the 'Annals of the Four Masters': as early as 1055 AD we read of a 'predatory excursion' which produced many 'spoils'. In 1314 AD marauding parties 'gathered herds, flocks and all valuable gear of the Corcamachs' from among 'Burren's uncouth ways, narrow gaps, crooked passes, rugged boulders and high sharp crests' (O'Donovan, 1851). In a reference from 1317AD we read of 'Burren's hilly grey expanse of jagged points and slippery steeps, nevertheless overflowing with milk and yielding luscious grass' (O'Grady, 1929). In 1600 AD, a raid by O'Donnell stripped the Burren of its 'cattle, flocks and booty', and later with 'enormous amount of cattle and plunder, they left the cleft stone passes of white Boireann behind' (Ó Cléirigh, cited in Ó Dálaigh, 1998).

Lemeneagh Castle in the south Burren. The original tower house is visible on the right, the main house a later addition, which boasts a very colourful history.

The agricultural significance of the Burren is also reflected in the fact that over twenty tower houses are found in the region (and fifty others in the adjoining baronies of Corcomroe and Inchiquin), several of which feature walled 'bawns' (from *ba dhúin*' or cattle enclosure), built to protect livestock. Many of these tower houses are located on the border of the Burren barony, leading historians to believe that they were built by the ruling O'Loughlin clan to defend the coveted winterage lands of the Burren. The famous Lemeneagh castle represents a fine example, dating from 1490AD, situated on the shale-limestone interface at the southern edge of the Burren. Dunguaire castle, located where the limestone of the Burren dips into Galway bay, is another good example.

Historical records also disprove the commonly held impression that the Burren is one homogenous block of barren limestone. As long ago as 1641 the enormous diversity of the Burren from an agricultural (and habitats) perspective was noted. The Book of Survey and Distribution (Simmington, 1641) recorded 35 different land types in the barony of the Burren, broken down into 121 sub-types based on their profitability. Fourteen different types of pasture, of 69 different profit levels, were distinguished, such as

Dunguaire Castle near Kinvara in the northern Burren.

Dwarfwood pasture (1/3 profit, 1/2 profit etc), Rockie pasture (1/4 profit,1/8 profit etc.), and Rockie shrubby pasture (1/3 profit etc.). This diversity, which is a hallmark of the Burren and a key contributory factor to the cultural and ecological wealth of the region, must be recognised and accommodated in management schemes today if these schemes are to prove successful in delivering upon their stated objectives.

Following the plantations of the 17th century when much the lands of the Burren were redistributed, sheep farming would appear to have attained prominence in the uplands, with vast flocks being kept by the main landowners, often on a year-round basis. The fascinating Statistical Survey of Clare (Dutton, 1808) records in the Burren that 'immense numbers [of sheep] are annually reared, and usually sold at the fair of Ballinasloe in October ... a small part feeds store bullocks'. Dutton also notes that 'the limestone crags of Burrin ... are, with some exceptions, devoted to the rearing of young cattle and sheep, and some so very rocky that four acres could not feed a sheep'. At this time for the smaller, local farmer however, mainly based in the coastal and valley areas of the Burren, a mixed economy based around the cow, pig, goat and the potato, on a dwindling land base, was prevalent.

Considering the contemporary perspective of the Burren as an ecological, cultural and geological haven, it is interesting to note that, historically, it was the agricultural quality of the Burren that took precedence, and seemed to generate most interest among observers. Ludlow (1651) refers to the 'sweet and nourishing' grass patches in the uplands; Dineley (1681) notes that the Burren 'raises earlier beef and mutton ... than any land in this kingdom, and much sweeter by reason of the sweet herbs intermixed and distributed elsewhere'; Lord Willies (1761) describes the sweet grass and shelter afforded to sheep by the uplands, so good that 'it fattens them prodigiously' producing 'near double the tallow of a sheep the same size fed upon rich pasture'; Roy (1788) claims that the hills provide 'the finest pasturage' for sheep; Coulter (1852) states that 'the fat sheep and cattle of the Burren' were 'proverbial amongst Irish agriculturalists' [all cited Ó Dálaigh, 1998]. Even the geologist F.J. Foot (1863) expounded the agricultural virtues of the Burren, in particular its soil 'than which none is more productive', providing pasture that is 'so rich and fattening'.

The Great Famine represented a watershed in terms of how the Burren was managed and how the landscape developed. Prior to the Famine, many upland areas continued to be used as extensive sheep walks, in stark contrast to cultivatable land which was mercilessly exploited by a desperate populace, increasingly dependent on the potato, cow and pig. The enormous pressure on the land prior to the Famine, when 400 people per square mile lived in areas such as Fanore and the Ballyvaughan and Turlough valleys, resulted by all accounts in a desolate, barren landscape. A record from 1852 by Coulter (cited in Ó Dálaigh, 1998) noted that a 'fuel famine' existed at this time in the Burren, such was the rarity of scrub, or woodland, with ferns, brambles and the stringy stems of mountain avens collected for fuel. Such a situation is almost unimaginable in the Burren today, and reminds us of how rapidly landscapes can evolve, over just a few generations in this case, as is clearly revealed in a comparison of images from either end of the 20th century.

This remarkable change in the landscape is a reflection of the fact that, following the ravages of the famine, human pressure on the Burren uplands has been diminishing from what would have been an unsustainable level of intensity. This relative lull in upland activity is reflected, as it was historically, in the gradual reversion of many upland grassland and pavement areas to scrub, primarily hazel.

Corkscrew Hill 1900. (Lawrence Collection, reproduced courtesy of the National Library).

To conclude, it is apparent even from this perfunctory review of the historical relationship between Burren farming communities and their environment, that the Burren is primarily a landscape of the people, their livestock, their traditions. This will come as no surprise to those who know the Burren well. Burren farmers, like the landscape itself, continue to evolve, in terms of their skills, technology, knowledge and ambition. We cannot expect these farmers, long characterised by their progressive and innovative dispositions, to ignore the contemporary realities that drive farmers elsewhere in Ireland and Europe. It is incumbent on us, the general public, to first of all understand this position and accommodate it, before we blindly berate transgressions from what we would consider appropriate or desirable management practices. Solutions with an agricultural *and* environmental relevance must be sought to address the challenges that this landscape faces today if these solutions are to prove sustainable in the long term, and if they are to meaningfully involve the farming community.

Corkscrew Hill, 2002.

4. Traditional Burren farming practices

The importance of traditional farming knowledge

In recent years, there have been numerous calls advocating a reversion to 'traditional' farming practices as a panacea for current environmental concerns, in the Burren and elsewhere. However, what exactly is meant by the term 'traditional' is subject to much conjecture and debate, as even within the Burren farming traditions change according to time and place. Perhaps the best description of traditional agriculture is that of Hopkins (1996), who claims that it involves a 'constancy of management practices, efficient utilisation of herbage and working with the inherent fertility of the land'. While a reversion to the traditional farming systems that once prevailed in the Burren would be mitigated against by contemporary market, labour and political factors, it is nonetheless worthwhile exploring in greater detail the traditional management of the Burren uplands, if only for the following reasons.

Traditions such as haymaking have become increasingly rare. Where these traditions survive, they have been greatly altered through mechanisation. Once hay would have been saved in cocks, later in 'square' bales, but now round bales are the fashion. The large size of these bales necessitates the use of heavy machinery for their distribution.

Firstly, certain elements of traditional management activities may well have relevance to the future management of Burren uplands – in terms of agriculture and, indeed, conservation. Such tried and trusted knowledge, hard-earned over many generations of intimate contact with the land, for example that relating to carrying capacities, breeds, grazing and feeding regimes, will be central to any informed future management strategy. After all, the heritage we treasure and hope to protect in the Burren today is little other than a reflection of the farming traditions of the past.

Disused stone farmhouses such as this would once have been teeming with life. Today only the shell remains, the feral goats in the background possibly the only connection with the lives and traditions that once inhabited this remote place.

Secondly, the Burren is a unique and challenging environment, one that has generated a distinctive agricultural and cultural response. Certain area-specific traditions, developed and honed over many generations, are being gradually eroded, and a valuable and intriguing slice of our cultural heritage is being lost forever. To quote the Burren-born poet John O'Donoghue (*Clare Champion* 11-2-00):

There is a world in the land, a farming world of the most sophisticated complexity and the most astute and rich memory that in the next ten years will have vanished completely. Isn't there something wrong with either our way of life or style of education that these huge ventricles of life, of memory or of perception are not being passed on?

A small sample of this vast store of management knowledge, and a reflection of its relevance, was garnered as part of this study, through informal interviews with some 25 elderly Burren farmers, along with an extensive literature review. The findings leave us in little doubt as to the range and diversity of traditional Burren upland management practices, and suggests that the common assumption that all of these areas were farmed homogeneously as winter pastures does an injustice to the hard work and ingenuity of these past generations of farmers. It also suggests that management recommendations for the future will need to far more subtle and discerning if this diversity in farming systems, and the rich natural and built heritage associated with them, is to be retained.

Yellow rattle.

Traditional farming systems in the Burren uplands

A fundamental point in relation to any analysis of Burren farming, traditional or current, is the enormous diversity of farm systems and practices present, a diversity that has its origins in the complex and highly variable geology, geomorphology and particularly soil characteristics of the uplands. Over many generations farmers have developed their own individual strategies to optimally exploit the agricultural potential of their particular holding. This has resulted in a great diversity of management histories that contributes to, and is reflected in, the unique cultural and ecological attributes of every upland parcel.

The agricultural potential of each and every upland parcel will depend on factors such as altitude, aspect, hydrology, soil characteristics, management history, accessibility, scrub and rock cover, among other things. This potential cannot be simply expressed in acreage terms as it is elsewhere, and was traditionally expressed in terms of 'the grazing of *so many* animals', e.g. the grazing of ten cattle. Upland parcels would have been rented and sold

There is a striking contrast between the agricultural productive capacity of the winterage in the foreground as opposed to that in the distance. The management of these lands, in terms of grazing regimes practised, would have evolved accordingly.

according to this potential, and not the acerage. This has been the case for many years: Dutton (1808) noted of the Burren that 'the greater part is let for low rents, often by the bulk, and not by the acre'.

How an upland parcel was managed also varied on an annual basis, dependent on markets, disease status, farm economy and, of course, weather conditions. Furthermore, the overall composition of any individual Burren farm would have been (and still is) critical in determining the management of units within it: the relative location of upland and lowland parcels, and the relative size of these parcels, are particularly important factors in dictating management. Little wonder such a diversity of management systems developed, contributing in no small way to the ecological diversity that evolved in tandem with them. This diversity of human utilisation and tradition must be protected.

In spite of this diversity in management systems, one significant distinction in terms of traditional upland management is worth noting – that between the farm system of the large, absentee 'rancher' farmer and his small-scale local counterpart (Table 1).

Table 1. A simplified summary of the different characteristics of the farm management systems practised by 'rancher' farmers and local 'smallholder' farmers in the Burren

Characteristic	Rancher farmer	Local farmer
Base	Outside Burren.	Locally based.
Farm size	Large, often with winterages rented in on an annual basis.	Relatively small, often renting out winterages to cover farm expenses.
Farm system	Specialised: cattle-based (18th and 19th cent. sheep-based).	Mixed: milk cows, beef cattle, sheep, goats, pigs, fowl and tillage.
Labour source	Herdsmen employed, paid in 'freedoms'. Also drovers.	Family labour. Very labour-intensive farm system.
Winterage start date	From October 9th (Kilfenora fair) onwards, stocked in stages.	Variable, depended on weather, land, market and financial conditions.
Winterage end date	May 1st on rented land, with a 2-week grace period for removal.	Again variable, depending on grass availability elsewhere.
Winterage stock type	Mainly older (2+) store cattle, prior to finishing.	Mixed livestock, mainly beef cattle in winter.
Stock origin	Brought from home farm or purchased at local fairs/off hoof.	Born on farm, or else bought as replacements for stock sold.
Summer use of winterage	Herdsmen's stock only.	Light use quite common, often involving sheep, dry cows, heifers etc
Feeding	Little or none.	Modest amounts, hay and root crops for cows, young or weak stock.

The smallholding farmer usually operated an integrated crop and livestock production system, quite labour intensive and designed to deliver a modest but steady supply of income through the year. These farmers were more likely to stock their upland parcels on a year-round basis, typically wintering cattle and summering sheep on them, while a parcel of goats were quite common, as were pigs, fowl and workhorses, as is revealed in the farm census data. Light supplementary feeding, particularly of young animals and in-calf cows, would have been common. In many cases, winterage parcels, which were always in great demand, were rented out by these smallholders to larger farmers for a fixed sum, often to cover the rent and rates due on the rest of the farm.

The large 'rancher' farmers in contrast owned or rented extensive Burren holdings purely as a winter store area for livestock (often sourced locally from the small farmers) prior to finishing them elsewhere on summer grass. Herdsmen would have been employed by these farmers to look after the stock, and drovers hired to transfer them, often via train. Due to the logistical difficulty involved, little feeding was administered and few livestock were kept in the uplands in summertime, save those of the herdsman.

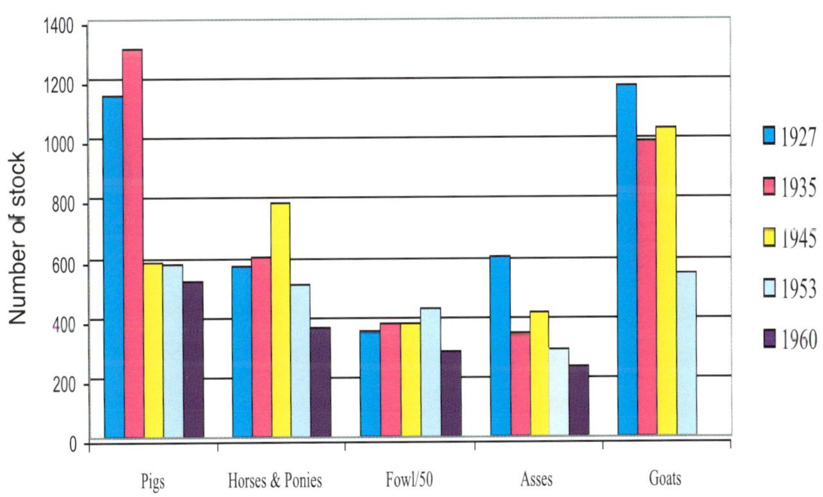

Total numbers of assorted stock types in Ballyvaughan RD, 1927-1960 (fowl numbers are divided by 50, source: CSO, assorted dates). This shows the prevalence of a traditional mixed farm economy in the Burren over the course of the 20th century.

This tradition of 'rancher' farming in Ireland has been well documented (Jones, 1983, 1995). Some characteristics of these rancher farmers would include their motivation 'by a sense of capitalistic enterprise', their 'lack of ancestral or customary ties to the land', and their emphasis on 'circulating', as opposed to 'fixed' capital. They operated on narrow profit margins and extensive scales, a strategy that 'impelled territorial expansion' (Jones,1995). Securing winter grazing rights on Burren uplands would have been a integral component of the systems operated by these farmers, some of whom were based as far away as the midlands and south.

The implications of these strikingly different upland management systems for current agri-environmental schemes in the uplands are enormous. The true 'winterage' tradition as it is commonly understood – extensive grazing limited to the October-April period with minimal external inputs - is, in a sense, more closely associated with these larger, externally-based, farmers, a few of whom continue in a similar vein today.

Local farmers however protest that these traditions are now being ascribed to them under schemes such as REPS, though their traditions in relation to upland management are in fact less absolute in terms of dates and management practices. This has led to a degree of conflict in the implementation of REPS, and has presented difficulties for some farmers who, under the scheme, have had to surrender their freedom to respond to unforeseen circumstances such as weather, disease and market conditions. More work needs to be done to resolve these conflicts if the scheme is to achieve broader acceptance and, ultimately, success in the region.

The Winterage tradition

The tradition of wintering for which the Burren is famed is thought to be quite ancient, and is based on sound agricultural principles, a fine example of the traditional practice of 'working with the inherent fertility of the land'. The limestone of the Burren hills is said to act like a giant storage heater, building up heat in summer and dissipating it in the winter months when the surrounding atmosphere is cooler. As a result, a relatively warm, dry lie is ensured for outwintering livestock, with the thin sparse soils scarcely masking the limestone's warmth, while also proving very resistant to waterlogging, muddying and erosion. The severity of the weather conditions is further mitigated by the wide availability of little hollows, scrub pockets and cliff faces for sheltering; where these were lacking, dedicated T-shaped (or cross-shaped) shelter walls were constructed.

Shelter wall, used to protect animals from the elements on more exposed slopes. Note the enriched (bright green) vegetation that has developed on this (leeward) side of the wall through years of dunging, urination and poaching.

Though nutritionally somewhat limited, the calcium rich vegetation and water of the Burren does promote strong animal bone growth, while the range of minerals provided by the diverse flora promotes animal health. The vegetation is quite resistant to excessive poaching, particularly the tough, resilient blue moor grass (known in Irish as *féar boirne* or burren grass) which dominates. The early spring growth of this April-flowering grass contributes to an extended grazing season, another reason why the uplands were so highly coveted for storing, if not fattening, animals over winter, at minimal cost, when very few alternative foddering options were otherwise available.

For many larger rancher farmers, winterages would traditionally have been stocked in stages as livestock became available, beginning with Kilfenora fair on October 9th. Locally based farmers would have been more responsive to weather and market conditions in terms of their winter grazing season. The winterage season normally extended to May 1st, though many farmers destocked their winterages in stages as grassland became available. In some cases cattle in good condition were removed early and sold before the forage quality deteriorated, and replaced by poorer, hungry animals.

Herd of suckler cows busily harvesting a patch of species-rich upland grassland in late Autumn, upholding a very ancient, and ecologially significant, agricultural tradition.

Significantly however, due to the limited availability of fertilisers, spring grass growth in lowland pastures was traditionally relatively late, and so a longer wintering season in the Burren uplands than that of today was commonplace. Livestock would, therefore, have been more likely to harvest even low-

nutritional value forage such as coarse grass, litter and scrub. Scrub is more palatable in springtime and early summer, as 'woody plants soon consolidate their summer growth by laying down secondary deposits of lignin, which makes stems inedible to stock from midsummer onwards' (Crofts and Jefferson, 1999). Thus, livestock browsing in the months of March, April and even May would have made an important contribution to limiting the spread of scrub.

The tendency to forage even low-nutritive vegetation would have been exacerbated by the fact that many winterages would traditionally have been overstocked (particularly on rented land) and little if any supplementary feed was administered. Extreme weight loss or even mortality among outwintering livestock was not unusual, particularly during the hungry months of March and April. Such a system would not be permissible today from either an animal welfare or an economic/market perspective. Traditionally, larger 'ranchers' were prepared to sacrifice a few animals, knowing that the good general health of the surviving majority ensured that they compensated at an accelerated rate when removed to fattening grass in summer. As a result, Burren-bred store cattle were always in strong demand, and earned a premium when brought to market, even following the introduction of the mart system in the early 1960's and the accurate weighing of these animals.

The cattle breeds used on winterages have evolved over the years. Back in the early 19th century the Longhorn - a hardy, hairy, dairy and beef animal - predominated, though a number of the 'Old Irish breed', a small black dairy animal, also remained (Dutton, 1808). The dual-purpose Shorthorn, first introduced to the region by the Burren Farming Society in the 1860s, became the dominant breed on the uplands for most of the next century. Even in 1960, Shorthorns represented 99% of cow breeds in Ballyvaughan RD, but were gradually replaced by Angus and Herefords for beef, and Holstein Fresians for dairying. Nowadays, continental breeds such as Charolais, Simmental and Limousine hold sway, driven primarily by consumer demand.

In terms of age, older 'store' cattle – anything from two to six years of age - were traditionally preferred on the uplands. At roughly two years of age, cattle develop their first two permanent incisors, adding two more every six months until a full complement of eight permanent teeth is reached roughly at three and a half years of age. Thus, older animals are more efficient at harvesting the tough, coarse grasses and shrubs found in the upland grasslands. In spite of this, by 1970 the importance of three-year-old-plus cattle in the Burren was diminishing in favour of younger (1.5-2.5 yr. old) animals (Finch, 1971), while today most Burren cattle are sold off as weanlings (under 1 yr. old).

The extent to which upland areas were traditionally grazed in summer is a subject of some contention today. Such was the importance of the wintering resource to the annual farm economy that, where possible, upland areas were closed off in summer to preserve them. Summer use was also discouraged by the very limited water availability, the excessive heat of the limestone, the relative unpalatability of the tough vegetation, and the danger of animals (particularly those in-heat or sensitive to warble flies) taking flight across the treacherous terrain during this season.

Sheep have always played an important role in grazing the Burren uplands and were, apparently, the dominant grazer in the 18th and early 19th centuries.

However it appears that the light summer use of winterages, particularly those locally owned and adjacent to the main holding, was far from unusual. Sheep were favoured for this task in many areas, particularly in the north Burren, due to their relatively high drought tolerance and their complementary grazing habits. Late calving or dry cows, young heifers and

other non-essential stock would also have been allowed to remain on during summer. In some cases uplands were grazed indiscriminately, on a year-round basis, particularly on those farms in the central Burren with insufficient lowland pasture, or on those upland parcels where water was freely available year-round and grass growth was sufficient to warrant (or even necessitate) such activity. Many herdsmen would also have retained their stock on the uplands in summertime, though in these cases the stock numbers involved would have been quite limited.

Other aspects of the traditional farm economy

Dairying was, until recently, an integral component of the local farm economy in the Burren, producing milk, beef stock and the dung that was badly needed for tillage. Dairy cows were rarely outwintered, their importance warranting them a more luxurious stay in the byre, from where their condition could be monitored and their dung collected. Back in 1808 Dutton noted that cows were 'generally kept in the house in winter, and fed with potatoes, usually boiled or mixed with bran'. The same source reveals that all excess milk at this time was churned into butter, packed into containers called firkins and sold at the market in Ennis for export via Limerick (Dutton, 1808).

Domestic butter and cheese making was common practice until centralised processing facilities became available. The mobile creamery was a regular visitor to the Burren in the early 1950s, extracting the cream from the milk and returning the remaining skimmed milk to the farmers for feeding calves and pigs. Deserted milk stands are a common sight in the Burren today, though dairying has also fallen victim to increased specialisation, and only a few dozen (highly productive and efficient) dairy farmers remain in operation.

Farmed goats were once very common in the Burren, and today's feral herds are a testament to this tradition. Often referred to as 'the poor man's cow', goats served several purposes in the small farm economy. They produced kids for sale at Easter, and butterfat-rich milk used for fattening young calves, domestic consumption, and for the treatment of asthma and eczema. Very importantly, these goats were actively managed, often through 'quiggering' (tying in pairs) to browse on, and control, scrub pockets. Old milking enclosures ('cahers') with annexed little huts called 'crós' are very common.

The crós were used to hold the kids by day until their mothers returned at night and they were allowed to suckle. The resultant 'milk kid' provided softer, sweeter meat than that of the equivalent 'grass kid', and was a traditional dish on St. Patrick's day and Easter Sunday.

Goat *cró* crudely made with rock (foreground, centre), with surrounding caher walls visible in the background. Such complexes are surprisingly common, even in quite remote areas, reflecting the traditional extent of goat husbandry in the uplands.

As historical records consistently reveal, sheep were also widely kept in the Burren. Their drought resistance and affinity for dry land made them an ideal grazer for upland areas in summertime. Dutton (1808) described the Burren as 'a sheep rearing country ... possessing a short bite, and requiring activity to gather it'. Lewis (1837) said of Carran parish that it 'is chiefly rocky pasture well adapted for sheep, of which the farmers' stock principally consists', while the upland pastures of Drumcreehy at this time were also 'principally devoted to the grazing of sheep'.

Sheepfolds, *puickets* (passageways through stone walls) and dipping tanks found in the uplands all attest to this longstanding tradition. In some parts of the northern Burren special 'sheep-cocks' were built in the fields for foddering sheep. These structures were usually built with branches for a base or 'butt', and had a long pole in the centre around which the hay cock was formed. This pole was used to prevent the cock from capsizing, and

smothering any sheep sheltering or feeding underneath. Galways were traditionally the main breed in the Burren, but now Suffolk and Texel breeds are more common.

A total of 48 shepherds were listed in Ballyvaughan RD in the 1901 census. Most of these shepherds lived in, or adjacent to, the shale areas around Mount Elva (16), Lisdoonvarna (7), Noughval (8) and Rathborney (9). These areas must have boasted extensive sheep flocks at this time, in much the same way as considerable expanses of the limestone uplands had herds of cattle, owned by larger rancher farmers. The high labour implicit in the management of sheep, along with the low prices for wool and erratic lamb prices, have seen sheep numbers steadily decline in recent times, especially on part-time farms.

Though it has been said of the Burren that 'the cowman rather than the ploughman is king', tillage was once widely practised in the Burren, particularly on the rich lowlands along the northern coast. Lewis (1837) wrote of Abbey parish that 'the greater portion is under tillage; the land along the coast produces good crops of wheat'. In the late 1960's a short-lived vegetable co op was founded in Ballyvaughan, and beet, potatoes and cereals are still grown in the vicinity. Further inland, even the smallest pockets of drift soils would once have been walled off for *gairdini* and cultivated by hand. On more extensive soil pockets, the imprint of old 'lazy beds', used for potato growing, are often visible. Traditionally, tillage was very labour intensive, and required the co-operation of neighbours. Many elderly farmers still fondly remember the *meitheals* and harvest dances that accompanied the thrashings of old.

Crop of oats in northern Burren, the upland hills in the distance. Today, tillage in the Burren is paractised by only a few farmers, mainly along the north coast.

Another interesting dimension of traditional management of the Burren is the many uses to which scrub, now commonly reviled as a threat, but once valued as a resource, (as is the case with goats), was put. Hazel would have been important for thatching, with metre-long rods called *scollops* used to pin down the thatch on a roof. Cutting cartloads of scollops for sale at local fairs was an important source of pocket money for many a Burren youth, along with the odd brace of snared rabbits. Hazel was also used for kindling, fencing, basket and furniture making. Lobster pots and calf muzzles were among the other interesting products generated. Winches, horses and axes were used to extract hazel trees by the root for feeding the ever-lighting kitchen fires of old, also serving to keep the land clear.

Finally, in terms of the traditional management of the Burren uplands, it is worth noting that controversial activities such as feeding, fertilisation and reclamation would all have been commonly practised, limited only by the high labour and limited technology involved. In terms of feeding, hay and root crops would have been sparingly administered to cows, sheep and young animals in wintertime when weather conditions dictated. Today, bagged silage is the main source of supplementary fodder used, very efficient in terms of its harvesting and distribution, and less dependent on the deteriorating summer weather conditions of late.

New (twin) members of the very extensive feral goat herds found in the Burren today.

Seaweed and farmyard manure were widely used prior to the advent of chemical fertilisers. Lewis (1837) noted that in Abbey parish 'sea manure is found in abundance on the shore' while in Gleninagh 'the portion in tillage is manured with sea-weed, an abundance of which is procured in the bay'. Upland areas were usually not fertilised due to their inaccessibility; fertilising was also said to 'soften' a winterage too much.

As regards reclamation, narrow pathways would have been cleared by hand to facilitate stock access, and pavement stone would have been mined as a ready source of wall building material. Patches of deeper soil would have been cleared of stones and tilled: the often-enormous mounds of picked stones that resulted from these clearances are still visible. Such labour intensive activity is no longer feasible. While mechanical solutions are now available, they are currently illegal in most of the Burren.

Examples of instances where limestone would have been removed by hand to meet the needs of stock access (left) and stock control (right).

Herding and fairs

Traditions relating to herding and herdsmen are particularly rich in the Burren, born of the historical penchant for rancher farmers to use the Burren as a winter storage area for cattle. Herders were employed by these farmers to supervise the livestock and the land, usually in return for the use of a house and gardens in the uplands, along with the 'freedom' to graze a defined number of livestock. Jones (1983) lists the main tasks of the herdsman as 'to search for stray cattle or sheep, to transfer the stock from one pasture to another, to drive them to and from market, and to distribute hay and fodder during winter months'. Other chores would have included keeping boundaries intact, cleaning watering points, and ensuring the safety and well being of their charges.

Herdsman's house and gardens in the south-west Burren.

Though money was rarely exchanged, no rates had to be paid, and so herding was considered to be potentially quite a lucrative profession. Jones (1995) claims that in the mid to late 19th century, 'herders were well paid, at least by the standards of income of Irish rural society', their small monetary remuneration 'significantly augmented by earning additional income (through saving crops, droving etc.) and through perquisites (income in kind, known as "freedoms") worth up to £50 or £60 per year'.

The 1901 Census offers an invaluable insight to the prevalence of herding in the Burren one hundred years ago. Some 103 herders were listed in Ballyvaughan RD, largely concentrated in the east Burren upland DEDs of Castletown (37), Carran (23), Oughtmama (16), Noughval (11) and Rathborney (7). Presumably, this would indicate that it was on the extensive upland limestone grassland holdings found in these DEDs that the true 'winterage' tradition predominated: in Castletown for instance, the number of indigenous farm families (16) was less than half the number of herders listed. Very few herders were listed from the more fertile coastal DEDs [Abbey (3), Drumcreehy (0), Gleninagh (2), Dereen (2)] where tillage and fishing would have sustained the indigenous small farm economy. Very few full-time herders remain today, though many of their successors now farm the lands on which their antecedents once herded, following redistribution by the land commission over the course of the 20th century.

Fairs were also an important element of the Burren farming tradition, a pivotal occasion in the annual farm calendar toward which the entire farm system was geared. The main fairs for Burren farmers were Kilfenora (October 9th and November 4th in particular), Gort, Ennis and even Ballinasloe. Other fairs and markets were held at Corofin, Ballyvaughan, Kinvara, Ennistymon and Turloughmore. Markets even further afield would once have been used: Lewis (1837), in describing Kilmoon parish in the south-west Burren, claims that it produced 'most luxuriant herbage, highly prized for grazing cattle, of which large droves are sent to the Cork and Liverpool markets'.

Epic week-long cattle droves to faraway fairs, undertaken by dedicated cattle 'drovers', accompanied by cart loads of hay, are still recalled. 'Stands' to rest the cattle at night were arranged en-route, while even for the local fairs, overnight droves were common. Similar long-distance livestock droves still occur in certain parts of Europe today, for instance in Spain, where traditional drove routes called *Cañadas* are used. Sometimes trains were used for long-distance cattle transfers ' to and from the Burren, but from the 1950s onward, lorries' became the main method used, as remains the case today. Most fairs died off in the 1960s, succeeded by the more equitable 'mart' system. Popular marts are now held at Ennis, Kilfenora and Gort, with evening sales a recent innovation in response to the increasing number of part-time farmers attending sales.

The future of tradition

It is important to remember that the Burren is not a wilderness or a strictly 'natural' landscape untouched by man, rather it is a 'semi-natural' or 'cultural' landscape, almost every nuance of which reflects the hand of man. Similar traditionally managed landscapes of high conservation value survive elsewhere in Europe but are becoming increasingly threatened. The extensive, low-input farming systems that created and characterised these areas are vulnerable in the face of the recent 'industrialisation of agriculture' that has led many such areas being 'improved' or abandoned. As in the Burren, particular species and communities of flora and fauna have developed in response to traditional management techniques, and any change in these regimes may have irreversible impacts on these species and their habitats. The impracticality and cost of their restoration only serves to accentuate the need for their conservation.

Admittedly, the relevance of many traditional systems is increasingly limited in a modern agricultural context by the high labour involved, the difficulty in marketing traditional produce, the erosion in available knowledge and the prevailing social and political climate. However these traditions still have a relevance, and will always have a value. The farming traditions that have developed in the Burren are in many ways as unique as the ecology, landscape and archaeology of the region, and in a modern context, are equally threatened, and equally worthy of protection. It is incumbent on us to find practical, feasible ways of supporting these traditions, so that future generations may not be disenfranchised of their right to enjoy this wonderful heritage

A hopeful future for the Burren's heritage?

5. Burren farming today

Times of change

The great political, technological, cultural and socio-economic changes that have taken place in Irish farming since the 1960's, and particularly since our accession to the EEC in 1973, have radically transformed the agricultural sector, and rendered many of the existing traditions and practices described earlier redundant. EU policy and international market demand have increasingly dictated what farmers produce, and new technology and a relentless drive towards efficiency and quality have greatly altered the methods of production. The historical dependency of rural communities on agriculture has been eroded, and increasing numbers of farm families have had to become reliant on off-farm income and direct payments from Brussels.

These forces of change evolve continually. The emphasis on production and efficiency that dominated Common Agricultural Policy (CAP) in the 1970s and 1980s has been gradually succeeded by policies that reflect the need for sustainability and environmental accountability. This very fundamental change in policy has been driven by disparate factors: international trade agreements, agricultural budget overspend and public indignation at environmental damage and overproduction costs. As a result, direct payments to farmers to deliver high quality agricultural *and* environmental goods are gradually replacing production-orientated price and market support systems. These significant changes seem certain to continue in the foreseeable future.

Increased levels of mechanisation have facilitated more labour efficient agricultural systems, but have also served to focus agricultural activity in the Burren on the more accessible and productive lowlands, upsetting the balance of land use that once existed.

Well-maintained stone walls are a reflection of the revolution in agricultural policies, whereby farmers are now paid to maintain and enhance the rural landscape.

Future changes in the CAP may include a further decoupling of direct payments from production, through the introduction of area-based payments, and a greater diversion of funds into rural development policies. Farmers are now being given the chance to reclaim their title as 'custodians of the countryside', and to be paid primarily to do just this. In the Burren, rather than constituting a threat, these changes may represent a tremendous opportunity for farmers and their local environment, particularly if we consider the wealth of heritage that the region supports, and the excellent track record that Burren farmers have in responding to changing circumstances and new challenges.

Undoubtedly, the Burren is no stranger to change – as we have seen, the history of Burren agriculture has been characterised by change. Historically, changes were gradually introduced (often in response to environmental change), and the environment readjusted accordingly, so that a degree of balance existed. In contrast, the unprecedented scale and speed of recent change has meant that the nature of the inevitable environmental response is far less predictable and more difficult to manage.

To attempt to register the extent of recent change and assess the current status of farming in the region, a detailed survey was conducted among 65 Burren farmers during late 1999 and early 2000. This survey successfully delivered information on farm family demography and economy, farm composition, general upland management issues and attitudes toward REPS. Very importantly, it also served to enlighten us as to the perceptions and decision-making processes of the all-important farmer on the ground.

A typical winterage scene in the Burren uplands.

Survey Results – farm family details

The average age of farm managers interviewed in the survey ('respondents') was 49.12 years, and the average number of farm dependents, including the farm manager and spouse (where present), was 3.28 per farm. Both figures are slightly below average for the mid-west region (Clare, Limerick and north Tipperary).

On over half (50.8%) of the farms surveyed, another source of earned income was being generated, with 44.6% of farm families totally dependent on farming, and the remaining 4.6% retired. If we consider the fact that, of those fully dependent on farming, many are elderly single farmers, these results indicate that the number of farmers involved in commercially viable full-time farming in the Burren is declining rapidly. This shift towards part-time farming is a national trend, and one of enormous significance in terms of how the Burren uplands will be managed in future, particularly as many of the management traditions that pertain to these uplands are relatively labour intensive.

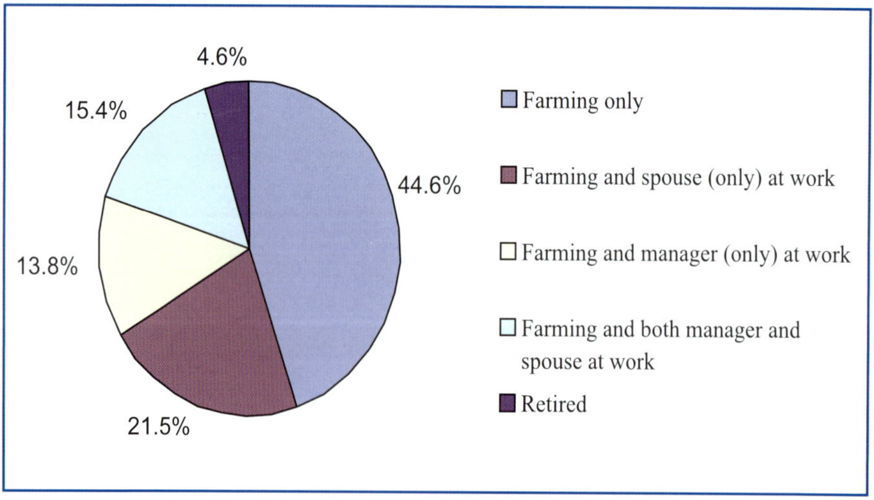

Farm family income sources in the Burren, 1999-2000 (source: Dunford, 2001).

These changes may be viewed as a response to declining farm viability, regarding which the future scenario appears quite grim. 43.1% of survey respondents felt that farming alone could not 'provide an acceptable living' for their family. An additional 44.6% felt that their farm was 'barely' able to provide enough for them, that any fall in market prices or income support would reduce their farm enterprise to insolvency. Again, of the 12.3% of respondents who felt that their farm was a 'viable' stand-alone enterprise, many were elderly widowed or single farmers whose opinions on viability are obviously very different and possibly unrealistic in today's world. On a more hopeful note, very few of the Burren farmers interviewed envisaged actually abandoning the land, with 83.1% seeing their future involving farming at some level, albeit, in many cases, as a part-time pursuit.

One of the greatest limits to enhancing farm viability cited by many Burren farmers is the high cost and limited availability of farmland, fuelled by demand from larger landowners, developers and investors. Considering the unique nature of Burren farming, and the individual nature of the management systems that exist within the region, it is highly desirable from a management perspective that these indigenous farm families should be encouraged to remain actively farming here, as most of them profess to desire. Changes in ownership usually signal a change in management, as the farm is subsumed into a larger entity or farmed by somebody who may be unfamiliar with the management history of the farm in question. This breakdown in ownership usually leads to a loss of traditional knowledge, hard earned knowledge often acquired over many generations of intimate contact with the land. The implications of this change in management for the natural and cultural heritage of these sites may only be guessed at.

In terms of diversifying and enhancing farm income, tourism has often been touted as the industry with the greatest potential for Burren farmers. Almost all (89%) of the farmers interviewed in the survey reported at least occasional use of their land by the general public. However, in stark contrast, a mere 15.3% of respondents claimed to benefit either directly, or indirectly (through a family member for instance), from tourism. This confirms that farmers are not among the prime beneficiaries of tourism, in spite of being the primary 'product' providers, and a big part of the overall attraction in their own right. A recent draft report (Consultative Committee on the Heritage of the Burren, 2001) concluded that, with regard to tourism in the Burren, landowners 'currently receive little direct benefit but bear a disproportionate part of the costs'.

The farmers of the Burren, and their traditions, constitute an invaluable cultural resource, an often-overlooked aspect of the Burren's rich heritage. This is reflected in the fact that the average period in charge of the farm among respondents was 25 years, over which time a considerable amount of knowledge must have been accumulated, in many cases informed by previous (on average three) generations of the family on the land. In any other profession, such a wealth of management experience would be invaluable: farming, particularly in an area such as the Burren, is no exception, and so every effort must be made to retain these farm families, and their traditions, on the land.

Can ways be found through which tourism can contribute to a secure future for the people of the Burren and their precious landscape?

Survey Results – farm composition

Survey results indicate that the average farm size owned in Ballyvaughan RD (which encompasses 75% of the Burren) was approximately 97.2ha (240 acres), up 8.6% from the 1991 CSO figure, with the average number of farms down by 8% to 299 over the same period. The average farm size in the broader Burren region was also found to be 240 acres, while the number of farms is estimated at approximately 378 (down 10.6% from 1991), though this number varies depending on where exactly we define the boundary of the area. The significant number of farmers based outside the region also complicates this approximation – farming sources claim that up to 700 farmers operate in the Burren.

Farm size in the Burren is thus, apparently, very high - 1.76 times the national average in 1991. However CSO data reveals that in Ballyvaughan RD, only half of the average farm acerage was classed as 'Agricultural Area Used' (AAU[2]) in 1991, and half of this again was classed as being useful only for rough grazing. Therefore, for the average farm in Ballyvaughan RD in 1991, the amount of 'good land' (suitable for fodder, crops and good pasture) was 25.4 ha per farm, more closely compatible with national (22.3 ha) and county (21.1 ha) averages. Survey results confirm the reality behind the farm size data: on the average Burren farm, almost two-thirds (60.6%) was classed by the respondent as 'upland' or 'crag', areas of very restricted productive capacity.

Average farm sizes in the Burren are quite high, as indicated by the low density of field walls. In some cases, entire townlands may lie within a single farm.

The variability of the agricultural productive capacity across the various DEDs of the Burren is evident if we look at the percentage of overall area in each DED that was classed as AAU in 1991 (Table 2). From this we see that, in 1991, only 51.44% of the area in Ballyvaugahan RD was listed as AAU, with roughly 70% of land in upland-dominated DEDs like Carran, Gleninagh and Oughtmama deemed not 'suitable' for agriculture, according to the definition used for AAU. In contrast, in the more fertile coastal DEDs such as Abbey and Dereen, roughly 90% of the land was classed as AAU.

Table 2 AAU in Ballyvaughan RD, 1991 (source: CSO, 1994).

DEDs	Total Area (ha)	AAU (ha)	AAU/Area (%)
Abbey	2516	2266	90.1
Carran	3561	1117	31.4
Castletown	2519	1377	54.7
Derreen	1891	1772	93.7
Drumcreehy	2100	947	45.1
Gleninagh	2863	850	29.7
Lisdoonvarna	1298	1012	78.0
Mountelva	2849	1293	45.4
Noughval	2608	1839	70.5
Oughtmama	3867	1196	30.9
Rathborney	2928	1247	42.6
Ballyvaughan RD	29000	14916	51.4

An interesting footnote to these figures on farm composition is the fragmentary nature of Burren farms. Only 56.6% (54.9 ha) of the average holding was situated adjacent to the farmhouse, according to survey results, while only 31% of farmers had their farm in a single unit. On average, farms were divided up into 2.82 separate parcels, the most distant of which was on average 6.4 km away, further complicating management.

[2] AAU - the combined area under crops, silage,hay, psture and rough grazing land in use (incluidng fallow and set aside land (CSO, 1994).

As the contrasting productive capacities of the land within this image testify, acerage is often a poor reflection of the quality of a particular farm. Low-lying areas tend to be quite fertile in contrast with the variable quality of upland areas.

Survey Results – farm system

Contrasting with the multifunctional systems of old (section 4), Burren farmers have become increasingly specialised of late, to the extent that almost 90% of farms now stock suckler cows, the average herd size being 30.8 cows. 46% of respondents stocked some sheep, on average 60 ewes. A higher than expected 18.5% of all respondents stocked dairy cows (average herd size of 31.9 cows), though some of these farmers would not have their dairy herds based in the Burren. 4.6% of all respondents stocked drystock (beef) only, in contrast to the former prevalence of this system. The significance of suckler cows to the economy of the Burren farmer, in contrast with his national, provincial or county counterpart, is clearly visible from 1991 census results (CSO, 1994).

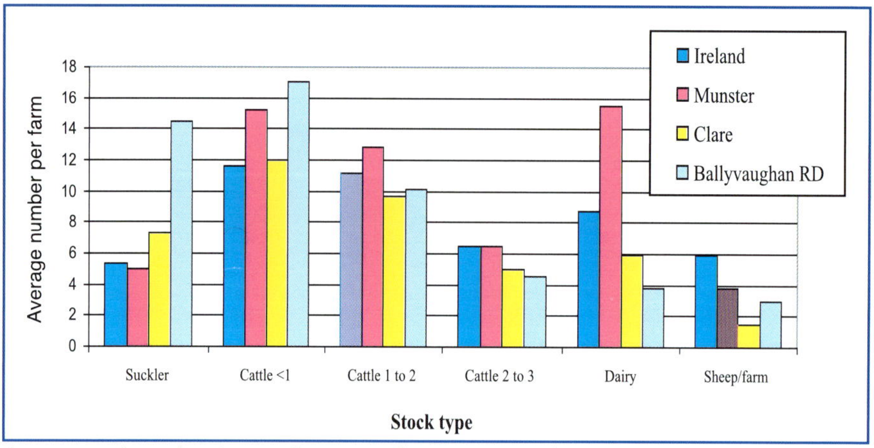

Average numbers of various stock types per farm at a national, provincial, county, and RD level in 1991 (source: CSO, 1994). NB: Sheep numbers are divided by 10.

Suckler cow based beef systems are now the dominant farm enterprise in the Burren. An interesting, and relative recent, development in this system has been the disposal of suckler progeny at a very young age. Over half (54%) of all suckler farmers sold all suckler calves before the end of their first year, while less than 7% reared any dry-stock to over 2 years of age, a system of cattle farming that was dominant in the Burren until quite recently. This trend is also reflected in the average numbers of beef cattle kept per farm by age group (over all farms surveyed): under one year, 30.1 animals; between one and two years of age, 11 animals; over two years; 31.9 animals.

Another interesting development is that the number of Livestock Units (LU) in Ballyvaughan RD, and thus the stocking rate, has shown a 39.1% increase from 1991-2000 according to survey results (35% increase in the Burren overall). Average stocking rates in Ballyvaughan RD stand at 0.65LU/ha, in the Burren 0.655LU/ha. These stocking rates, though apparently rising very rapidly, are nonetheless quite low, confirmation of the limited agricultural potential of much of the land in the Burren.

If stocking rates are calculated according to AAU, the low stocking rates of the Burren are still apparent, though to a lesser extent. At a national level the rate is 1.47 LU/ha (AAU), for county Clare it is 1.19 LU/ha (AAU), for Ballyvaughan RD it is 0.91 LU/ha (AAU). The corresponding rate for the wider Burren region would be 0.94 LU/ha (1991). The changes in stocking levels between 1970 and 1991 at a DED level in Ballyvaughan RD demonstrate a steady overall increase in stocking rates over this period. Again, upland DEDs such as Carran and Gleninagh have the lowest stocking rates.

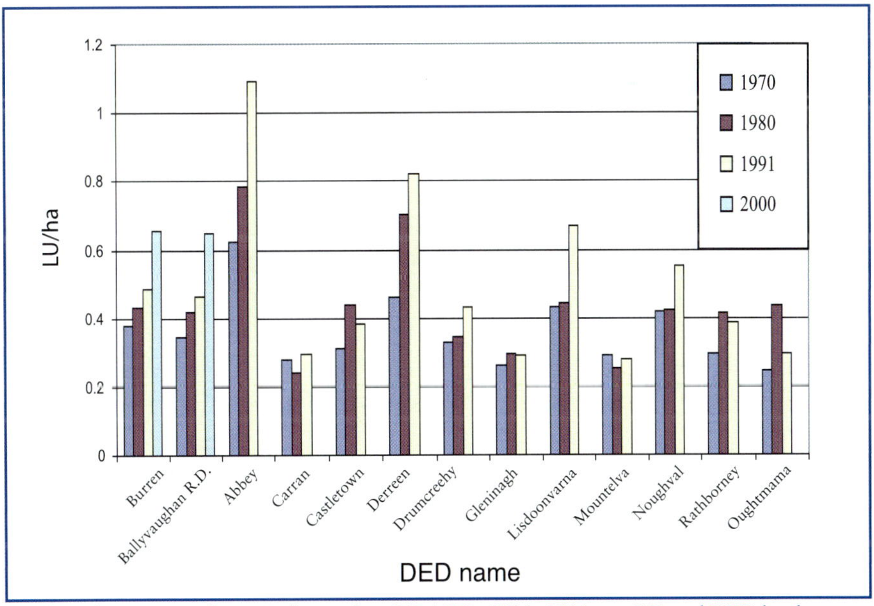

Stocking densities (LU/ha) in Ballyvaughan RD 1970, 1980, 1991 at a RD and DED level (source: CSO, 1994). Also included are estimates for the 'Burren' region and Ballyvaughan RD in the year 2000 (source: Dunford, 2001).

Breeds are also changing, according to survey results: Charolais bulls were found in 31% of herds, while Limousine (21.6%) and Simmental (10.8%) were the next most popular breeds, a clear response to market demand for higher quality beef cuts. Some farmers feel that these breeds are inherently less suited to the rough terrain and coarse herbage of the Burren uplands than breeds such as Hereford, Shorthorn and Angus that dominated formerly, and thus they require greater care and supplementary nutrition.

To accommodate these increased stocking levels and new nutritional demands, and to circumvent the unpredictable weather conditions of late, round bales of silage are now produced on 77% of Burren farms. The average silage crop was approximately 281 bales (or equivalent). Only 21.5% of respondents used silage 'pits', most preferring the mobility and general convenience afforded by the round bale. Less than one in five farmers (18.5%) now cut hay, though very few, if any, are wholly dependent on it.

Ironically, the impact of the higher stocking densities and specialised livestock systems in the Burren region has actually been to contribute to the marginalisation and neglect of many winterages. The increased productive

For many Burren farmers, the demands of off farm work, the need to improve efficiency and quality, and demands for improved pollution control has prompted them to invest in slatted units and baled silage as an alternative wintering option

capacity of lowland areas following reclamation in the 1980s and the subsequent increase in slurry and fertiliser application, allied with more efficient fodder harvesting and slatted housing, seems to have absorbed much of the increase in stock. Simply put, the relatively greater control, convenience and nutritional quality provided by the use of slatted houses and/or supplementary feedstuffs is steadily displacing upland winterage grasslands as a wintering option, as time-pressed farmers struggle to streamline their farm operation. While most uplands are still being stocked, in general the extent to which these uplands are being grazed has declined considerably, presenting some serious environmental challenges (section 6).

Survey results – upland management

The average winterage size among survey respondents was estimated at 61.5ha, and the average stocking rate of these winterages, or parcels thereof, was estimated by respondents (in 47 cases) at 0.56 LU/ha over a six-month winterage period, or 0.28 LU/ha annually. This accords very well with recommended annual rate of 0.25LU/ha for the 'conservation grazing' of lowland calcareous grasslands recommended by Crofts and Jefferson (1999). On average, winterages were grazed for 4.6 months.

Over half (55.4%) of all respondents claimed to use some upland grasslands for summer grazing, with the average acreage involved estimated at 27.5 ha. This represents 27.2% of the overall acreage owned by these particular farmers. The extent to which these grasslands are used varies widely. In many cases such grasslands have been summer-grazed for many years, often a reflection of the imbalance between the upland and lowland components of the particular farm involved. As vegetation studies reveal (Section 6), heavy summer grazing over time usually results in the loss of native flora.

Of the 40% of respondents who claimed to feed animals outdoors on their winterages with silage, the average feeding period was 12 weeks, and the average fodder amount was 189 bales, or approximately 2.24 bales per day. The numbers of stock involved could not be estimated with a reasonable degree of accuracy in most cases. Again, this practice has significant ecological and environmental repercussions, not least of which would be the reduced level of foraging by relatively well-fed livestock, leading to the under-utilisation of the grazing resource and a consequent loss in plant species diversity.

On average, some 2190 Euro was spent annually on chemical fertiliser by each farmer interviewed, corresponding to roughly 14 tons per farm, below average for the mid-west region. Some 18.5% of farmers stated that they commonly fertilised 'upland' areas, often by hand. Fertilising is said to 'sweeten' winterages, making them more productive and suitable for summer grazing, though also making them 'softer' - less durable and susceptible to poaching by livestock.

Again, as with summer grazing and feeding, the ecological impact of fertilising upland grasslands can potentially be very negative (section 6), which is why restrictions on these three activities have been incorporated into the Burren REPS agreement. Research elsewhere has indicated that fertiliser restrictions do not necessarily compromise production, and can actually result in considerable savings. The restrictions appear to be effective in the Burren: in the survey it was found that non- participants in REPS spent 1.8 times as much as REPS participants annually on fertiliser, a welcome vindication of the scheme's effectiveness, and one consistent with national trends.

Visible ecological change at a former feeding site: the impact of additional nutrients, disturbance and weed species creates a distinctly different flora.

Survey Results – scrub

One of the more visible implications of recent changes in upland management has been a gradual encroachment of scrub in many areas. According to two thirds of farmers interviewed, scrub constitutes a problem on their holding. Hazel is considered to be the main threat, particularly on upland areas, while whitethorn and blackthorn are more prevalent on areas where deeper soil pockets exist. From a farmer's perspective, the factors contributing to scrub encroachment include the abandonment of goat farming, general reductions in upland grazing pressure, REPS limitations on upland grazing, the lack of manpower, deteriorating weather conditions, and lack of uses/markets for scrub. The likely reason is in fact a combination of these and other factors.

Scrub affects all aspects of the Burren. From a farmer's perspective, the grazing potential, and thus economic value, of the land is diminished, while stock management, herding, and general access for livestock is rendered more difficult and dangerous. Many archaeological, historical and traditional cultural sites are being damaged: structurally in some cases, in other cases subsumed and forgotten, leading perhaps to their unintended destruction by future generations.

Scrub growth resulting in the severe disturbance of a court tomb structure.

Hazel scrub (and in the background, ash trees) encroachment onto an area of bare pavement, north Burren. Even such barren areas, where not overly exposed to the elements, are susceptible to scrub incursion.

Floristically rich grasslands and pavements (designated under the EU Habitats Directive as a priority habitat) are steadily being replaced by species-poor grassland, scrub and woodland. The prized and distinctive geology and geomorphology of the Burren is also being compromised. Elsewhere in Europe, the loss of limestone grasslands to scrub is an issue high on the conservation agenda, with some countries forced to initiate expensive restoration schemes that are invariably limited in their effectiveness.

Though experiments have been initiated by Dúchas and the IFA into scrub removal, the prohibitive cost of clearing even small areas of established scrub suggests that the emphasis must be urgently placed on the prevention of further scrub encroachment. That this is achieved in a manner that accords well with, and does not impinge unduly upon, the needs of local farmers is central to the sustainability of such an initiative. Teagasc are currently spearheading a research programme to establish the suitability and effectiveness of various methods of achieving these objectives.

REPS

REPS, the Rural Environmental Protection Scheme, is the main policy instrument responsible for delivering sustainable agriculture in Ireland, and was introduced in 1994 under EEC Regulation 2078/92. Its objectives include establishing environmentally friendly farming practices, the protection of wildlife habitats and endangered species of flora and fauna, and the production of quality food in an extensive and environmentally sustainable manner. The unique circumstances under which Burren farmers operate were acknowledged through the compilation of a special set of 'Conditions for the Conservation of the Burren to be Applied under REPS' (Department of Agriculture and Food, 1995). These conditions were agreed by a multi-party working group in November 1995, and included measures to limit the extent of summer grazing and supplementary feeding on upland grasslands.

Of the farm managers interviewed, some 70.8% were REPS participants, double the county average. The inherently extensive nature of farming in the region, the large farm size and availability of supplementary payments for NHA lands, as well as the uniquely tailored Burren agreement, all contributed to this high membership level. REPS participants interviewed were found to have lower overall stocking densities, fertiliser bills and silage production levels than their non-REPS counterparts. The scheme was also found to be an important factor in contributing to the viability and survival of some smaller farm enterprises which would otherwise struggle to survive. REPS has also generated considerable spin-off employment in rural areas, often for part-time farmers, particularly in activities such as stonemasonry, hedge cutting and fencing.

Stone wall maintenance, fencing and farmyard repairs were the main entry works required of respondents who joined the Burren scheme, while the main day-to-day changes resulting from membership included stock management adjustments, tidying up, and alterations to the feeding and fertilising regime. Over half (56%) of respondents said that the changes in their farm systems as a result of REPS membership were slight, 24% claimed that major changes had to be made, while a further 20% felt that they had to make no significant change.

Though certain measures are unpopular, in general the scheme is viewed quite favourably by the farming community, with 94% of respondents considering it to be a positive development. The main benefits cited were the improved tidiness of the countryside, in particular field walls and farmyards. However, many respondents would like to see the restrictions on grazing and feeding times relaxed, and in general more discretion and flexibility to be allowed to the farmer to cope with the unexpected, particularly weather and disease conditions on the farm. Some 13% of survey respondents had been penalised, mostly for shortcomings in boundary wall maintenance. Concern was also expressed in some quarters that the scheme was contributing to the encroachment of scrub and poorer farming practices.

In spite of these negative sentiments, REPS is, and will continue to be, of great relevance to the livelihood of Burren farmers and the protection of their heritage. In an area where the proactive co-operation of farmers in the conservation process in essential, the popularity of the scheme offers real hope. With continual honing of the scheme's measures to effectively address the main environmental concerns, while simultaneously engaging with the often-unforeseen reality of day-to-day farming, the future effectiveness of the scheme can, and will, improve upon an already promising start.

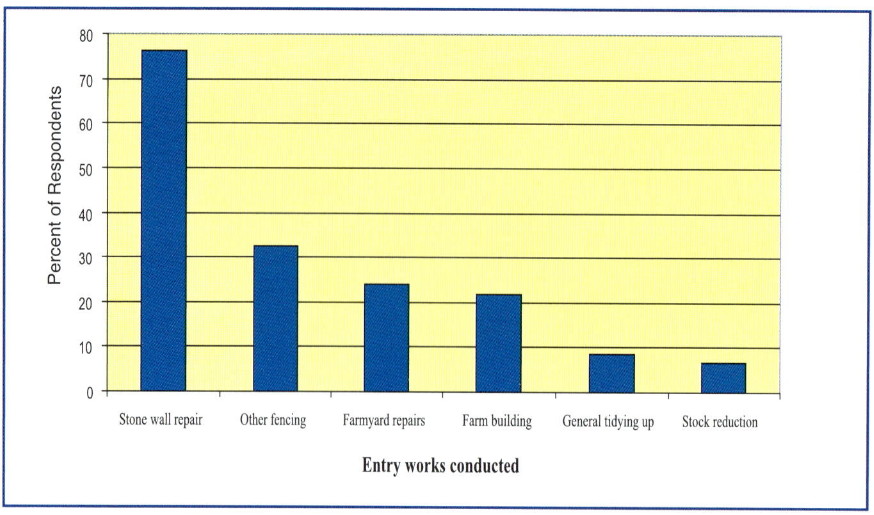

Main works required of farmers upon entry into REPS (source: Dunford, 2001)

Conclusions

The current situation of farming in the Burren is scarcely recognisable compared to that of thirty years ago. From being the dominant rural activity of a few decades ago, less than half of all farm families now rely on farming alone for their income. Farms are being continually consolidated to form larger holdings, with the lower manpower available increasingly concentrated on mechanically accessible lowland areas, while uplands are simultaneously marginalised. Markets and direct payments have heralded the dominance of suckler cows on the uplands, their nutritional requirements increasingly met by imported fodder to the neglect of the winterage vegetation.

Suckler cow and weanling, representing the predominant agricultural system practised in the Burren uplands today.

These changes look set to continue. When survey respondents were asked what major changes (limited to two responses) in Burren farming they anticipated over the course of the next decade, over half (58%) predicted a significant increase in part-time farming. 52% of respondents felt that there was going to be an increasing consolidation of farms, with big farms increasing in size at the expense of the smaller, less-viable operations. The opportunity to react to these changes, which are already well underway, in a sensible and effective manner, is greatly strengthened by the increasingly positive attitude to nature conservation on the part of local farmers.

We also learn from this survey of Burren agriculture that the major period of change initiated prior to, but accelerated by, accession to the EEC in 1973, and exacerbated by the recent growth in the national economy, is set to continue. Attitudes are changing rapidly, and the ideas that prevailed prior to this period of unprecedented change are dying with the last generation to have practised them. A new generation of farmers has arrived: well educated and enterprising in areas beyond the realm of agriculture, technologically adept, operationally efficient, economically aware, pragmatic in their attitude to 'the land', not bound by loyalty to the ideals of family and community as were their predecessors, worldly-wise, ambitious and well informed.

For these farmers, success is no longer determined exclusively by farm production, as there is now access to many other sources of income. Traditional farming skills such as the ability to judge or treat stock or to

Mixed livestock grazing, north Burren.

cultivate land, while still important, are less relevant than the ability to position the farm operation so as to optimally benefit from the range of compensatory schemes available. While many of these changes may be construed as potentially threatening to the farming traditions of the Burren, from a strategic management perspective it has become increasingly easier to manipulate future developments in agriculture, and thus conservation, due to the unprecedented dependency on external stimuli, particularly the CAP. While formerly, these external stimuli may have contributed to environmental damage through reclamation and fertilisation, in future they may well have a more beneficial impact. That this is done in a sustainable fashion that benefits this area and its people is the challenge we face, and to brave this challenge we must be as informed as possible regarding the relationship between farming and this precious environment.

Small blue butterfly on bracken plant.

6. The ecological impact of agricultural practices

Introduction

Burren grasslands are notoriously complex and difficult to describe scientifically, due to the high degree of variation in soil and vegetation characteristics over very short distances. Visibly different vegetation types - heaths, orchid-rich limestone grasslands and fertile pastures - frequently intermix, usually reflecting the variability of the local soil resource. While the seminal work of Ivimey-Cook and Proctor (1966), among others, described the vegetation of the Burren in some detail, little work has been done to investigate the impact of management practices on grassland composition.

Such a study, though complicated by the range and diversity of the ecological and management characteristics of upland grasslands, is vital to inform future management strategies. Consequently, as part of this research project, the vegetation, soil, general environmental and management characteristics of 214 Burren uplands sites were described in detail. These sites were located within over 130 different land parcels spread across 13 DEDs in the three RDs of the Burren. All flowering plants, grasses, sedges, shrubs and mosses encountered, and their percentage cover values within a $1m^2$ area called a quadrat were recorded at each of five points at the site in question, then averaged to give an overall reading for the site.

Example of a typical $1m^2$ vegetation quadrat taken on an upland grassland.

The management of each grassland patch sampled was assigned to one of five broad categories for simplicity: undergrazed[3], winter grazed, winter grazed with additional light summer grazing, heavily summer grazed (or grazed on a year-round basis), and sites used for feeding. Altitude, aspect, soil characteristics and other relevant information was also noted. This information - taken over two seasons (1999-2000) - was assembled in a digital database and then analysed using dedicated computer programmes. Some of the findings generated are now discussed.

Thyme broomrape, another Burren speciality. The absence of any green pigment in this plant reflects the fact that it does not photosynthesise, thus no chlorophyll is produced. Instead this plant is parasitic on the roots of wild thyme, hence its name.

[3]Sites were classed as undergrazed based on vegetation condition, litter (dead plant material), and disturbance levels, information from farm managers, etc. Undergrazed grassland sites on which scrub had encroached were not sampled

Results – general grassland types

One of the most common methods of analysing vegetation data is through a programme called TWINSPAN (see Kent and Coker, 1992). When this programme was used to analyse the entire vegetation dataset collected in the Burren, the 214 samples were repeatedly divided until such time as eight sub-groups were distinguished, solely on the basis of their vegetation composition. Each sub-group generated represents a different type of upland grassland, and the samples within each group have more in common with each other (in terms of their vegetation composition) than they would have with the samples from any other group. These groups (for simplicity referred to as A to H) are now briefly described, in terms of their vegetation, management regime and main environmental characteristics.

The first group ('A') was composed of vegetation samples containing high levels of weed species, clovers and agricultural grasses. Further analysis revealed that most of these samples were taken from heavily disturbed feeding sites, with many of the species found having been most likely introduced in supplementary feedstuffs and boosted through the high levels of dunging and urination. Close to a feeding site the damage is extreme, while moving further out the effect is gradually diminished and spatially variable. This vegetation, having been badly damaged, is of limited conservation value.

Typical site from which a 'Group A' vegetation sample might have been taken. Note that the severity of the impact diminishes as we move away from the feeder (bottom left).

Typical (group B) vegetation found on an upland grassland that has been traditionally grazed on a year-round basis, and possibly fertilised. Bracken and thistles are common.

The vegetation samples in the second group ('B') were found to be associated almost exclusively with grasslands that were grazed on a year round basis. The average number of plants found in quadrats taken at these sites was almost 14% below average, and the vegetation was similar to that found in lowland grasslands, with a lot of agricultural grasses, and herbs such as common sorrel, ribwort plantain, clovers, daisies, buttercups, dandelions and thistles. Most of the interesting Burren flora that might have existed under a more sympathetic grazing regime had been lost at these sites, diminishing their conservation value. This emphasises just how important a factor management can be in determining vegetation composition, and highlights the ecological significance of the Burren's winterage system relative to summer bias of systems that prevail elsewhere.

A third group ('C') of vegetation samples was found to be associated with what are known locally as 'strong' winterages: i.e. upland areas with a good depth of soil and rich (often rank) vegetation that are mainly winter grazed and unimproved. These grasslands sport a rich and diverse flora, with a strong element of grasses and common herbs like clover, ribwort plantain, common knapweed and yarrow, but also with a number of typical Burren species such as bloody crane's-bill, ladies' bedstraw, and eyebright species. Highly coveted agriculturally, these grasslands are also very important from a nature conservation perspective.

The type of vegetation found in group C samples is often very colourful.

A fourth group ('D') was found to contain samples from grasslands that had been moderately disturbed through feeding and poaching, but also a number of samples taken from undisturbed species-rich grasslands on deepish mineral soils. The samples associated with feeding sites were usually taken at a distance (>60m) from the feeder, with species such as the common daisy, oxeye daisy and mouse-ear hawkweeds typical of such sites. These species were also found in the high-diversity grassland samples, but in addition a large proportion of distinctive species such as the Burren rock-rose, cat's foot, and common centaury were also present. This undisturbed species-rich vegetation type (*Antennaria-Hieracium*) has been identified in the Burren as being of very significant conservation importance (Ivimey-Cook and Proctor, 1966, Keane, 1990).

A fifth group ('E') was found to be closely associated with upland grasslands of moderate quality (average soil depth) that are lightly summer grazed. The flora as a result is an interesting mixture of native Burren species (blue moor grass, ladies' bedstraw, wild thyme, harebell, devil's bit scabious) with a significant additional 'weed' presence (buttercup, daisy, dandelion). Plant

In group D samples, some of the characteristic species would include the burren rock-rose and mouse-ear hawkweed, seen here with blue moor grass.

Group E vegetation is an interesting mix of typical Burren species such as harebell and ladies' bedstraw, and introduced species such as red clover and cat's-ear.

species diversity as a result was found to be slightly above average, and though the flora is not as intact as might be preferred, such grasslands are nonetheless very important from an ecological perspective.

Group F: Like group E, this grassland is dominated by blue moor grass, but tends not to be grazed as well, often contributing to a degree of scrub incursion and the predominance of species tolerant of low grazing levels.

A sixth group ('F') was distinguished: though similar to group E samples in terms of their moderate productive capacity, the signs of disturbance were greatly reduced as these grasslands are normally winter-grazed only. Some of these grasslands were somewhat undergrazed, as was reflected in the relatively high levels of scrub species. The flora is dominated by species such as tormentil, bloody crane's-bill, bird's-foot-trefoil, vetches and ladies' bedstraw, species that are able to clamber over the often-high vegetation present. Again, this vegetation would be of considerable conservation value.

A seventh group ('G') was composed of samples taken from poor quality grasslands on very thin soils. The strong presence of species such as mountain avens, cat's foot, carline thistle, wood sage and goldenrod form a distinctive vegetation which reflects the poverty of the soil resource, as these species usually compete poorly on deeper soils. A good number of these sites had been lightly summer grazed, a fact that may have contributed to the introduction of some weed species such as buttercups and dandelions. These weed species are, however, unlikely to pose a threat due to the difficult growing conditions on these most characteristic of Burren grasslands.

Dryas grassland with cat's foot and bird's-foot-trefoil, typical of Group G grasslands, which often contain a distinct additional ruderal element.

Group H vegetation is a well-developed limestone heath, in certain cases sporting characteristic species such as bell heather (pink) and slender St. John's wort (yellow).

A final group ('H') was identified, quite similar to the previous group G in terms of the distinctive presence of species such as mountain avens, heathers and slender St. John's wort. However, the imported weed element of the previous group was not present, leading to a lower species diversity, a difference that may be linked to the relatively low levels of summer grazing and disturbance on these exposed, elevated sites. This group would represent a more intact or undisturbed version of the latter group, again representing one of the most characteristic and significant Burren grassland communities, sometimes described as a limestone heath.

From this analysis, we see that there are several distinct grassland 'types' to be found in the Burren uplands. In some cases it is clear that the management of these sites is the primary factor responsible for the difference in vegetation composition. This is particularly evident in the case of groups A and B which represent, in a sense, management extremes, as is reflected in their heavily disturbed and radically different flora. Management factors, particularly the extent of summer use, would also appear to be important in distinguishing between certain groups (e.g.: groups E and F, G and H).

Aside from the extremes of group A and B, soil depth was found to be the critical factor in determining vegetation composition at a site. On relatively deep soils (e.g. group C), mesotrophic *Cynosurus cristatus-Centaurea nigra* grasslands predominate. In contrast, on very thin soils (e.g. groups G, H) we invariably find the heathy *Dryas octopetala-Carex flacca* community (nomenclature follows Rodwell, 1992). All along the 'vegetation continuum' that exists between the extremes of rich mesotrophic grassland and depauperate limestone heath are found innumerable intermediate combinations of *Sesleria*-dominated calcareous grasslands (e.g. groups D, E, F).

In general, we can summarise from these vegetation studies that the composition of any individual grassland patch will be a reflection of the juxtaposition of natural factors (particularly soil depth but also altitude), and management history (particularly the degree of summer grazing or feeding practised).

Dryas dominated limestone heath on a thin rendzina soil, one of the most characteristic vegetation communities found in the Burren uplands. Soil characteristics are a critical factor in determining vegetation composition.

The impact of management on plant species diversity

While the above analyses give us some idea about the influence of various factors on vegetation *composition*, when we simply compare the *diversity* of plant species found under different management regimes, some interesting trends are also noted (Table 3).

Table 3. Variations in plant species richness (per $1m^2$ quadrat) under different management regimes (± standard deviations).

	Overall	Under Grazed	Wintered	Light Summer Graze	Heavy Summer Graze	Feeders
No. of samples	n=214	n=23	n=100	n=49	n=16	n=26
Species per m^2	28±4.9	23.03±4.0	28.9±4.4	29.2±4.1	23.1±3.9	29.5±5.3
Deviation (%)	-	-17.76	+3.39	+4.29	-17.50	+5.36

The most characteristic Burren rose, the burnet rose, in flower.

Of the 1,070 quadrats described (214 sites, 5 quadrats per site), on average 28 different species were found per square metre, with up to 45 species in some cases. This is again a testament to the remarkable richness of the Burren flora: a typical perennial rye-grass sward elsewhere in Ireland would probably contain less than a half dozen species. However, in Table 3 we see that at the extremes of very low levels of grazing ('undergrazed sites') or very high levels of summer (year round) grazing, plant species diversity levels are significantly reduced, by 17.8% and 17.5% respectively.

In the case of the undergrazed sites, grasses such as blue-moor grass and fescues, along with some competitive herb species and litter (decayed vegetation, particularly *Sesleria* or blue moor grass) eventually become dominant, making it very difficult for weaker members of the native flora to survive, thus reducing diversity and producing a dull, lifeless grassland. While in some cases these grasslands may be succeeded by a scrub community, they may also continue to exist as low-diversity open grasslands, particularly in exposed situations. Thus, the impact of undergrazing in the uplands is not manifested solely in terms of scrub encroachment as is often mistakenly thought – extensive areas of species-poor undergrazed grasslands with little or no scrub cover are also surprisingly common.

Undergrazed upland grassland, dominated by blue moor grass and litter.

On summer-grazed sites the increased levels of grazing, dunging and urinating create an enriched habitat more suited to competitive and ruderal species that are typically found on lowland grassland. Over time the native flora is ousted, unable to compete, and diversity gradually drops off. Only on the occasional, inaccessible rocky oasis within these relatively homogenous grasslands does the original flora survive, testament to the degree of management-induced change that has occurred.

The modest increase in plant species richness associated with feeding and light summer grazing is due in large part to the introduction of grass and weed species in addition to the native flora present. This highlights the fact that species diversity data does not always accurately reflect the conservation status of a grassland. Overall, from a management perspective, it may be concluded that a regime based exclusively or primarily on the practice of winter grazing appears to be the optimal option, upholding diversity while minimising disturbance of the most characteristic plant communities.

The impact of management on grassland composition

Another reflection of the influence of management on vegetation composition may be had by looking at the response of individual species to different management regimes. The frequencies of occurrence of some of the most characteristic Burren species (blue moor grass, eyebright species, yellow rattle, mountain avens and goldenrod) are shown under three levels of grazing pressure. It is clear that increased levels of summer grazing, particularly heavy summer grazing, reduces the levels of these species. Similar trends were found for some of the Burren's orchid species.

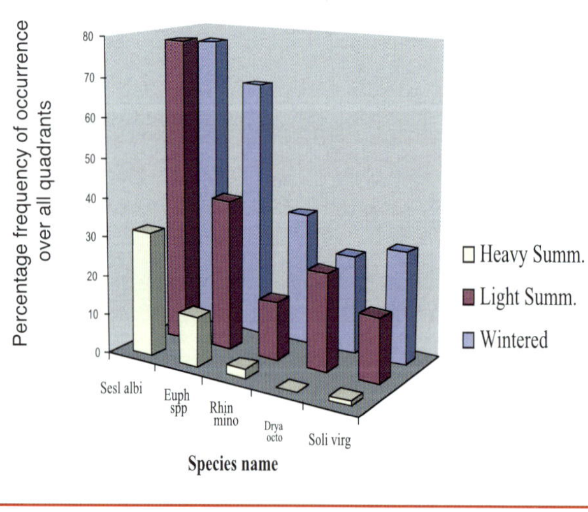

Percentage frequency of occurrence (per quadrat) of some dry limestone grassland species under three different grazing regimes.

In contrast, the frequencies of occurrence of some common lowland grassland species (crested dog's-tail grass, mouse-ear chickweed, speedwell, dandelion, common daisy) clearly increase with increased levels of summer use.

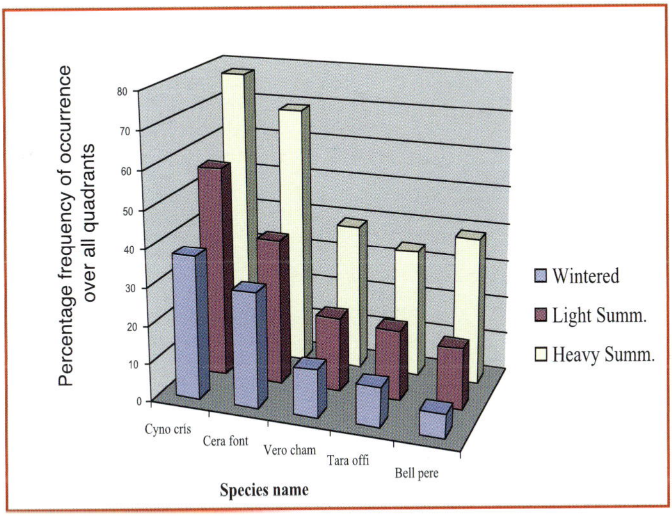

Percentage frequency of occurrence (per quadrat) of some mesotrophic grassland species under three different grazing regimes.

In terms of general vegetation cover, management also has a clear impact (Table 4). We see that increased levels of summer grazing serve to increase grass and moss cover at the expense of herb/forb and scrub cover. On undergrazed sites grass and scrub cover is also increased at the expense of the herb element. Feeding sites provide the highest levels of herbaceous flora, though much of this flora would be composed of transient ruderal species. Otherwise, it is the traditional regime of winter grazing that supports the flora with the highest percentage cover of forbs. It is primarily to this diverse herb flora that the Burren owes its international distinction, and it is this flora that we must work to protect.

Table 4 Average percentage cover (per transect) of functional vegetation types under varying management regimes (± standard deviations).

	Grass	Forbs	Mosses	Sedges	Scrub
Overall Average	32.28 ± 11.4	53.63 ± 13.2	4.5 ± 4.3	3.67 ± 3.7	1.94 ± 7.3
Undergrazed	39.45 ± 15.3	45.7 ± 14	3.78 ± 3.5	4.83 ± 5.8	4.64 ±13.1
Wintered	30.1 ± 11.2	56.54 ± 12	4.2 ± 3.4	3.93 ± 3	1.76 ± 6.5
Wintered and lightly summered	33.54 ± 12.5	50.6 ± 12.5	4.82 ± 3.5	4.27 ± 4.2	1.3 ± 4.1
Heavily Summered	41.09 ± 16	42.84 ± 12.6	8.25 ± 6.8	1.53 ± 2.8	0.68 ± 2.8
Feeding sites	26.53 ± 9.7	61.51 ± 13.2	3.5 ± 3.5	1.91 ± 2.1	2.3 ± 8

Other findings

The impact of mixed livestock (cattle and sheep) grazing on vegetation composition was also investigated. Despite the complexities involved, results tentatively indicated that grazing systems that involved the use of sheep resulted in large increases in the occurrence of plants such as perennial rye-grass, crested dog's-tail, selfheal and daisies, compared with cattle grazed sites. The frequency of occurrence of hazel and orchid species appeared to suffer from sheep grazing, as did overall diversity levels, as grass cover increased and forb cover diminished. The proclivity of sheep for grazing the nutritive flowering heads of orchids and other flowering species has not enhanced their reputation among ecologists as conservation grazers on calcareous grasslands.

Typical burren assemblege of blue moor grass, squinancywort and wild thyme, all closely associated with areas that have a thin soil cover.

Fertiliser was found to drastically reduce diversity on the small number of sites where its use was known to occur, with competitive grass and flowering plant species such as ribwort plantain, clover and daisy greatly increased in terms of their frequency of occurrence. The fact that fertiliser use is normally associated with summer use of uplands makes its specific effect difficult to distinguish. It was however found that high phosphate (P) levels were associated with mesotrophic grassland species, while the more characteristic Burren flora was found at lower P levels, another indicator of how potentially damaging fertilisation would be on these species-rich grasslands.

Analysis of soil samples indicated that pH levels fell as altitude increased, most likely due to the higher leaching levels present. Also pH was found to be lower on deeper soils as the influence of the calcareous bedrock in balancing the acidifying effects of leaching is reduced. Also of interest is the association between typical Burren species such as blue moor grass, wild thyme, and squinancywort with shallow soils, confirmation of the importance of the region's unusual soil attributes in sustaining this unique flora.

Greater knapweed and oxeye daisy.

Conclusions

Though fraught with complexities, a simple analysis of the findings from this ecological survey reveals that the management regime under which a grassland patch is managed may have a very significant impact on the vegetation composition of that patch. In particular, this research confirms the vital ecological importance of the winter grazing regime, and its close association with the most characteristic grassland communities and species found in the Burren. Should this tradition cease, and these grasslands were to be grazed on a year round basis as elsewhere, then it is clear that not only would the plant species diversity be diminished, but that the vast proportion of the native Burren flora would be lost at these sites. Similarly, the disturbance and enrichment associated with feeding sites severely damages the native flora. These findings offer some justification for the limitations placed on the further extension of heavy summer grazing and supplementary feeding on upland grasslands under the REP Scheme.

The other significant finding from this analysis is the serious loss in plant species diversity that results from the undergrazing and neglect of these grasslands. The significance of this threat is exacerbated by findings from the farm survey (section 5) which suggest that many upland areas will be consigned in future to such neglect. The extensive nature of this threat, and the implications for the agricultural, ecological and cultural wealth of the Burren, suggests that such a discontinuation of agricultural activity is perhaps the greatest future threat to the species-rich upland grasslands of the Burren.

The main challenges of future management systems will be to find a feasible solution to the management extremes of year round grazing (foreground) and neglect or abandonment (background).

7. The future: threats and promise

Summary

The Burren has been aptly described as '150 square miles of paradoxes' (Robinson, 1999). These paradoxes represent a challenge to the researcher and policymaker alike, but are an important element in the attraction that this unique place holds for people, and something to be celebrated. Some of these paradoxes and complexities have been explored in this study, and some interesting findings have been generated. However this study is only a small (but hopefully pragmatic) step in the long road towards a fuller understanding of what is a wonderfully complex and diverse landscape.

The singular beauty of the Burren landscape, as seen at Mullaghmore.

In this study we have seen what a longstanding and profound impact farming communities past and present have had on the geomorphological, cultural and ecological evolution of the Burren landscape. We have learned about their unique and innovative farming systems, structures and techniques. We have read of the high quality of the produce generated by these farmers, widely renowned over many centuries. The rich, diverse legacy of tradition, structure and folklore left by innumerable generations of Burren farmers represents a vital, threatened component of the region's heritage, urgently in need of recognition from a utilitarian and cultural perspective. For the Burren is primarily about farmers - past, present and, we sincerely hope, future.

The upheaval of the past three decades or more in the agricultural sphere have upset the fortuitously harmonious agriculture-environment relationship that defined the Burren and its people, and left us with a daunting challenge for the future. Specialisation, intensification and marginalisation have occurred within farm systems of late as farmers have struggled to reposition themselves in the shifting sands of today's global agricultural model. Increasing levels of off-farm employment and changes in ownership are also contributing to changes in management. These management changes have, in turn, resulted in significant environmental change, not least from an ecological perspective, as vegetation studies conducted in this study have revealed.

Turning the clock back to the extensive, low-input, labour-intensive systems of old that upheld the rich diversity of the uplands is not a realistic option for the majority of Burren farmers. Instead, a new dynamic is required which will explore the common ground between the best of the old and new, in order that farmers continue to make a decent living in a way that continues to protect and enhance the landscape, while securing the invaluable heritage of the Burren for future generations.

The typically rich, diverse and colourful grassland flora of the Burren – with bloody crane's-bill (purple), bird's-foot-trefoil (yellow) and milkwort (blue).

Management recommendations

In terms of specific management recommendations for upland grasslands generated by this study, these can be surmised quite simply as those practices that accord best with the core 'traditional' principles of high levels of winter grazing and minimal external input. Ecological studies clearly indicate that the greatest current management threats to the Burren's floral diversity come from a corruption of these traditional principles through undergrazing, extensive long-term summer grazing, fertilisation and excessive supplementary feeding.

All available evidence suggests that upland grasslands are set to become increasingly marginalised and underused, as a diminishing number of farmers focus their limited labour resources on more fertile and accessible lowland sites. It is likely therefore that the threat posed to the biodiversity of the Burren through undergrazing will dominate the conservation agenda in the Burren for some time to come, as is already the case on many limestone grasslands in the UK and elsewhere (Grayson, 2000). The striking lesson to be learnt from the experience of other countries is that a proactive approach

While some areas have suffered severe scrub encroachment, the Burren still retains extensive open grassland and pavement areas. If this balance is to be maintained however, a concerted effort will urgently need to be made by all involved.

to limestone grassland management is infinitely preferable to the pursuit of expensive, controversial, reactive solutions that are inherently limited in their effectiveness.

While certain areas in the Burren have suffered extreme levels of scrub encroachment, our priorities in terms of management must lie with the retention of the species-rich limestone grasslands that remain. The core point is this: though we may disagree as to the nature and extent of scrub encroachment as a management issue, we must address this issue now if we are to be in a position to manage the Burren uplands at all in future. As experiments have shown, the rate of encroachment can be relatively fast. The best time to effectively address the future management of the Burren uplands is right now.

To help offset the undergrazing of uplands, grazing periods, as they are currently defined under REPS (October 1st to April 30th), need to be fully utilised. Animal numbers should be maintained or, where deemed necessary and feasible, increased. Ideally, there should be minimal recourse to supplementation, at least until the winterage has been thoroughly grazed. The judicious use of concentrated supplementary feed, instead of the liberal

Exclosure erected in the Burren National Park by Bonham (1980), showing the impact of a 22 year cessation of grazing by large herbivores on upland vegetation.

Have feral goats a role to play in the future management of the Burren?

use of silage, offers the possibility of enhancing grazing levels on uplands, rather than restricting them through substitution, as currently happens. The sustainability of this option is greatly enhanced by the fact that it has an increasing agricultural relevance, particularly in the prevailing scenario of rising silage production costs and the drive for efficiency and convenience among the large numbers of farmers with off-farm employment. Teagasc is currently investigating the agricultural and environmental implications presented by such a change in management.

The practice of summer grazing (sometimes involving fertilisation), which, in contrast, involves the attempted over-exploitation of upland grasslands, is somewhat more complicated. In many cases this is a traditional practice, often a response to limitations in available summer land. While heavy summer grazing over a long period does destroy the native plant communities, light summer use may have a neutral, or sometimes positive, impact on floral diversity. However, to define appropriate or acceptable levels of summer use is very complicated, due primarily to the inherent natural and management diversity of each and every upland parcel. While REPS measures to limit the further extension of summer grazing onto upland grasslands may be justified, the 100% penalty currently imposed for breaches of this seasonal grazing REPS prescription is unduly harsh in light of the limited damage caused by short periods of summer use.

In terms of the type of grazing animals to be used in the uplands, barring a radical change in policy it is likely that the suckler cow, and in a few cases beef store, will remain the dominant types, as they currently represent the optimal economic choice. From a conservation perspective, the presence of these animals, if managed appropriately, may maintain or enhance diversity through grazing, defecating and occasional poaching. The use of low-maintenance 'traditional' breeds may also need to be investigated, to assess the advantages (lower maintenance) offered by their hardier natures, though 'exotic' traditional breeds with little market value are unlikely to prove a sustainable alternative for grassland management under current market conditions.

Sheep appear to represent too much of a threat ecologically, and an excessive burden from the management perspective, to warrant intervention in their current decline. Again, if 'conservation grazing' is to work in the long term in the Burren, the methods adopted should, ideally, be justifiable from an agricultural perspective. Goats may be useful for scrub control, as experiments elsewhere in Europe testify (Bohnsac and Carrucane, 1999), though the cost and complexity of their management for such a purpose would mitigate against their widespread use.

The lengthy tradition of sheep farming in the Burren is in decline.

The enormous ecological benefits of the winter grazing regime are clear. However, it must be accepted that, in spite of our best efforts, 'even in grasslands of low fertility … winter grazing will usually still allow scrub to encroach …a regime based on winter grazing will therefore usually need to include provision for regularly repeated scrub clearance to remove the gradual accretion of woody plants throughout the site' (Crofts and Jefferson, 1999). This scrub clearance was effected traditionally using goats or by hand harvesting of scrub, practices which are no longer realistic due to the high labour and limited return involved.

Thus, restorative management may be necessary on upland grasslands that have succumbed to scrub encroachment. Selective scrub removal and chemical treatment to facilitate livestock access to pastures, watering and feeding points may be needed at certain sites. In less extreme cases, an increase in grazing pressure may be required to remove the build up of rank vegetation, litter and scrub, and return the grassland to a favourable conservation status. Traditional breeds, a change to concentrate feeds and an extended grazing season are all possible solutions to be investigated in this regard.

These animals, and how they are managed, have an important role to play in the future evolution of the Burren landscape, and in particular the maintenance of its species-rich limestone grasslands.

A reduction in the amount of silage administered will probably be a key factor in future grassland conservation initiatives in the Burren. As is recommended practice for conservation grazing schemes in the UK, if supplementary nutrition is required, concentrates or even hay should be used, administered sparingly in sacrificial low-diversity locations once the winterage has been well grazed, and the animal then allowed forage to appetite (Crofts and Jefferson, 1999).

One final recommendation, possibly the most important, must be made. Due to the bewildering inherent natural diversity of the Burren uplands, combined with the unique management history of each parcel, the only effective way of assessing the management needs of the Burren is at the level of the farm, or better still the individual parcels within it. Prescribing general rules, dates and stocking levels at a regional level may be justified from an administration perspective, but is a foolhardy exercise from an environmental and agricultural perspective, prone to cause as much inadvertent damage as good.

Farmers are undeniably the best placed and best informed to make such decisions, but they must understand what they are protecting, and it must be fully justified to them as to why they are protecting it. Put simply, farmers are *the* experts in terms of land management in the Burren, but in an age and place where environmental concerns are increasingly taking precedence over the agricultural produce generated, farmers will need to re-evaluate their farm systems and re-deploy their expertise accordingly.

Perhaps agri-environmental schemes of the future will make the proactive shift of putting the onus on farmers to deliver the environmental goods using the skill and innovation that has always been their forte. This is a tall order, one that would require a great degree of education, trust and responsibility, and a fundamental re-evaluation of the role of Burren farmers. However, to continue to do otherwise might fatally compromise the inherent diversity of management systems that currently survive, and that should be valued and protected as much as other elements of the Burren's heritage.

General recommendations and conclusions

We are currently in the midst of a period of great change, challenge and opportunity in the Burren. It is a critical time from a management perspective, as the decisions we make together now will impact on the evolution of this landscape for many years to come. The timing is, in a sense, fortuitous: the political and financial wherewithal appears to be in place, there is a significant body of expertise and experience available to inform our decisions, and farmers are increasingly willing to co-operate. This is an opportunity not to be missed, an opportunity to deliver a model of sustainable agriculture for the Burren, one that meets the needs of farmers and their environment, while benefiting from, and facilitating the needs of, other sectors such as tourism.

To fully avail of this opportunity, it must first of all be appreciated that the Burren is a dynamic, living landscape closely attuned to human activity, enlivened by human presence, threatened by human neglect and excess. To protect and manage such a landscape will require a greater degree of co-operation and understanding among all stakeholders in the region, and an appreciation of their ostensibly conflicting, but ultimately compatible, perspectives. That we proceed in an informed manner will require an extensive, ongoing programme of research and education. For its part, Teagasc is deeply committed to the sustainable agricultural management of this most precious of national resources, and is currently sponsoring a number of 'conservation management' research initiatives aimed at securing the livelihood of Burren farmers in a manner consistent with the area's environmental needs.

The continued absence of an integrated education programme that is relevant, insightful and accessible, is a major obstacle to real progress in the sustainable management of the Burren. There remains a critical lack of awareness among many farmers of the resource that they are being paid to protect, of its significance and of the subtleties involved in its management. Equally, many representatives of local, regional and state management bodies, visitors and even well intentioned 'conservationists', remain hopelessly oblivious to the important role that farmers play in protecting, and contributing to, the Burren's heritage, and of the constraints within which these farmers operate. This is a situation that needlessly stifles real co-operation and development.

A late summer favourite, the harebell, in flower.

REPS is the main policy measure to implement meaningful change at a landscape scale in the Burren. Its high uptake in the region suggests that a well-designed and administered scheme can offer considerable potential in this regard. REPS undoubtedly plays an important role in the economic support of farmers at a time of diminishing incomes, and works to limit some of the more damaging management practices. While REPS has been criticised in some quarters as having failed to deliver, it must be borne in mind that through REPS, for the first time ever, farmers are being asked to move beyond their production-orientated mentality and embrace measures that give conservation of the environment precedence over agricultural production. Such a radical shift in perspective will surely take time to sink in, and we must be patient and understanding while it does.

REPS is quintessentially a scheme which depends on the positive, proactive co-operation and enthusiasm of farmers for its ultimate success. There is a strong case to be made that REPS should allow farmers to operate under defined 'limits of tolerance' within which they can be more progressive in exercising their knowledge and discretion in responding to climate, market and other conditions. To facilitate this, a more equitable reward system should also, perhaps, be considered, through which farmers are proactively

paid to deliver environmental goods, rather than reactively penalised for digressions from predetermined, standardised, and often inappropriate, criteria.

New opportunities will have to be sourced for Burren farmers so that they can continue to survive on the land, as farming alone can not and will not provide sufficient income for many of them. Tourism, frequently mooted as the panacea for the financial concerns of Burren farmers, has so far failed to deliver on its enormous potential in the region, as reflected in the dramatic gap between the percentage of farmers who host tourists on their land, and those who benefit in any way. At best, tourism-related activity will only appeal to a limited number of farmers: for the remainder, the provision of information, education and skills will be more relevant. The new Teagasc Opportunities programme is a good first step in this regard, but will require the co-operation and resources of other bodies to really deliver in the longer term. New enterprises and marketing opportunities will also need to be explored'

As this research has confirmed, it is imperative that the Burren farming community be brought into the conservation process, informed, equipped and encouraged to lead the charge in terms of proactive conservation initiatives. Burren farmers have a vast store of experience, and an appetite to participate in the future management of the area, according to survey results. As the main stakeholders in the Burren, farmers have the most to gain and the most to lose. Elsewhere in Europe, the interconnectedness of agriculture and heritage conservation has been recognised (Pienkowski, 1998) and supported by governments and organisations such as the European Forum for Nature Conservation and Pastoralism (www.efncp.org). It is essential therefore that Burren farmers bring their vision and leadership to the management process, and thereby become (in the words of the Burren IFA) 'architects of their own destiny'.

Future strategy

The Burren is not a landscape frozen in time: it is an evolving entity, subjected to and influenced by various, continually evolving, forces. The same can be said of Burren farmers. We cannot expect that they should adhere to traditional, low-impact systems and ignore the sweeping political, market and technological changes that propel them towards a more efficient, profit-orientated system of agriculture.

We cannot manage the Burren's expansive and diverse landscape in isolation from the farmers, and their traditions, that have influenced it so strongly and for so long. What we must do is to acknowledge the overwhelmingly positive role played by these farmers, and support them – through research, education, and income – to ensure that this critical role is upheld and strengthened into the future. Farmers for their part must deliver, and deliver in a proactive fashion, fulfilling this role and thus their potential as custodians of the countryside, and meeting the attendant challenges head on, as we know they are singularly capable of doing. Only then will the high levels of public expectation and continued financial support be justified, and the future of the Burren and its people guaranteed.

In terms of meeting these objectives, this study is but the beginning, a basis from which to move forward in a more informed, inclusive and ultimately, effective and sustainable manner. All parties involved in the day to day management of the Burren must take up this new challenge, re-evaluate their role, leave the baggage of the past behind, and look to the future with renewed hope and vigour.

Where do we go from here?

Acknowledgements

The author would like to thank the following individuals for their assistance in the compilation of this book:

Jim Flanagan, Director, Teagasc, 19 Sandymount Avenue, Dublin 4.
Donal Carey, Director of Operations, Teagasc North, Grange, Co. Meath.
Michael Miley, Head, Public Relations & Information, 19 Sandymount Avenue, Dublin 4.
Tom Shanahan, CAO, Teagasc, Ennis, Co. Clare.
Liam Fitzgerald, Drystock Specialist, Teagasc, Athenry, Co. Galway.
Sean Regan, Environmental Specialist, Teagasc, Athenry, Co. Galway.
Dr. John Finn, Environmental Specialist, Teagasc, Johnstown Castle, Co. Wexford.
Dr. Con O'Rourke, Research Development, Teagasc, 19 Sandymount Avenue, Dublin 4.
Jane Kavanagh, Publications Manager, Teagasc, 19 Sandymount Avenue, Dublin 4.
Liam Jones, Teagasc (Retired Advisor), Ennis, Co. Clare.
The administration and advisory staff of Teagasc in County Clare.

The role played by former Teagasc Director Dr. Liam Downey, former Teagasc CAO in Clare, Mr. Michael McGrath, and former advisor at Teagasc Ennis, Co. Clare, John Keane, in initiating and supporting this research project is also gratefully acknowledged.

Thanks also to:
Dr. John Feehan, Dept. of Environmental Resource Management, UCD.
Dr. Grace O'Donovan, Dept. of Environmental Resource Management, UCD.
Ute Bohnsac, Burren Action Group.
Michael MacMahon, The Heritage Council.
Gerard Kerin, Chairman, Clare IFA.
Michael Davoren, Chairman, Burren IFA.
All members of the Burren IFA group.

Sincere gratitude to the Burren farming community, for their hospitality, generosity and support, in particular to the following, for their wonderful stories of Burren farming in times gone by: Miko Beakey (Aughavinnaun), Mary Brogan (Cappagh), Jack Casey (Murrough), Tommy Fahy (Finnevarra), Joe Gilligan (Glencolumkille), Paddy Healy (Baur), Jack Hynes (Murtyclough), Miko Hynes (Croughville), Christy Irwin (Ballyconroy), Paul Keane, RIP (Commons North), Gerard McGann (Tullycommon), Pete Neachtan (Funshin), P.J. O'Halloran (Knockanes), Sean O'Halloran (Noughval), P.J. Purcell (Munnia), and Sonny Quinn (Ballymurphy).

Finally, to Ann, Caoimhe and Alannah for their patience, love and support.

All photographs and text by Brendan Dunford BE, MSc, PhD.
For more information, please visit www.burrenbeo.com

burrenbeo.com

Plant species referred to in the text

English Name *Latin name* Irish name (after Nelson, 2000)
Ash *Fraxinus excelsior* Fuinseog
Bee orchid *Ophrys apifera* Magairlín na mbeach
Bell heather *Erica cinerea* Fraoch cloigíneach
Bird's-foot-trefoil *Lotus corniculatus* Crobh éin
Blackthorn *Prunus spinosa* Draighean
Bloody crane's-bill *Geranium sanguineum* Crobh dearg
Blue moor grass *Sesleria albicans* Féar boirne
Bracken *Pteridium aquilinium* Raithneach mhór
Burren rock-rose *Helianthemum canum* Grianrós liath
Buttercups *Ranunculus spp*. Fearbán
Carline thistle *Carlina vulgaris* Feochadán mín
Cat's foot *Antennaria dioica* Catluibh
Cat's-ear *Hypochaeris radicata* Cluas chait
Clovers *Trifolium* spp. Seamair
Common centaury *Centaurium erythraea* Dréimire Mhuire
Common daisy *Bellis perennis* Nóinín
Common heather *Calluna vulgaris* Fraoch mór
Common knapweed *Centaurea nigra* Mínscoth
Common sorrel *Rumex acetosa* Samhadh bó
Crested dog's-tail grass *Cynosurus cristatus* Coinfhéar
Dandelions *Taraxacum* spp. Caisearbhán
Devil's bit scabious *Succisa pratensis* Odhrach bhallach
Early Purple Orchid *Orchis mascula* Magairlín meidhreach
Eyebrights *Euphrasia* spp. Glanrosc
Fescue grasses *Festuca* spp. Feisciú
Glaucous sedge *Carex flacca* Cíb liathghorm
Goldenrod *Solidago virgaurea* Slat óir
Grass of Parnassus *Parnassia palustris* Fionnscoth
Greater knapweed *Centaurea scabiosa* Mínscoth mhór
Harebell *Campanula rotundifolia* Méaracán gorm
Hazel *Corylus avellana* Coll
Ladies' bedstraw *Galium verum* Boladh cnis
Milkwort *Polygala vulgaris* Lus an bhainne
Mountain avens *Dryas octopetala* Leaithín
Mouse-ear chickweed *Cerastium fontanum* Cluas luchóige
Mouse-ear hawkweed *Hieracium pilosella* Searbh na muc
Oxeye daisy *Leucanthemum vulgare* Nóinín mór
Perennial rye-grass *Lolium perenne* Seagalach buan

Plant species referred to in the text

Ribwort plantain *Plantago lanceolata* Slánlus
Selfheal *Prunella vulgaris* Duán ceannchosach
Slender St. Johns wort *Hypericum pulchrum* Beathnua baineann
Speedwell *Veronica* spp. Lus cré
Spring gentian *Gentiana verna* Ceadharlach Bealtaine
Squinancywort *Asperula cynanchica* Lus na haincise
Thistles *Cirsium* spp. Feochadán
Tormentil *Potentilla erecta* Néalfartach
Vetches *Vicia* spp. Peasair
Whitethorn *Crataegus monogyna* Sceach gheal
Wild thyme *Thymus polytrichus* Tím chreige
Wood sage *Teucrium scorodonia* Iúr sléibhe
Yarrow *Achillea millefolium* Athair thalún
Yellow rattle *Rhinantus minor* Gliográn

Bibliography

Bohnsac, U. and P. Carrucane (1999). *An Assessment of Farming Prescriptions under the Rural Environmental Protection Scheme in the Uplands of the Burren Karstic Region, Co. Clare.* Prepared for the Heritage Council. Kilkenny: The Heritage Council.

Brady Shipman Martin (Natural Environment Consultants) (1996). *Burren National Park Study: Draft strategy for North Clare area.* Dublin: The Stationary Office.

Consultative Committee on the Heritage of the Burren (2001). Draft report to the Minister of Arts, Heritage, Gaeltacht and the Islands. Unpublished.

Cotter, C. (1999). Western Forts Project: Cahercommaun Fort, Co. Clare - a reassessment of its cultural context. *Discovery Programme Reports* 5, 41-96. Dublin: Royal Irish Academy/Discovery Programme.

Crofts, A. and R.G. Jefferson (1999). The Lowland Grassland Management Handbook (2nd Edition). English Nature/The Wildlife Trusts.

CSO (Central Statistics Office) (1994). Census of Agriculture, June 1991. Detailed Results. Dublin: Central Statistics Office.

CSO (Central Statistics Office) (1998). Census of Population, 1996 – Small Area Population Statistics – Phase 1 and 2. Assorted DEDs. Dublin: CSO.

D'Arcy, G. and J. Hayward (1992). *The Natural History of the Burren.* London: Immel Publishing.

Department of Agriculture and Food (1995). *Conditions for the Conservation of the Burren to be Applied under the Rural Environmental Protection Scheme.* Circular 84/95.

de Valera, R and S. O'Nuallain (1961). *Survey of the Megalithic tombs of Ireland Vol. 1. County Clare.* Dublin: The Stationary Office.

Drew, D. (1990). The Hydrology of the Burren, County Clare. *Irish Geography* 23(2), 69-89. Dublin: Geographical Society of Ireland.

Drew, D. and E. Magee (1994). Environmental implications of land reclamation in the Burren, Co. Clare: a preliminary analysis. Irish Geography 27(2), 81-96. Dublin: Geographical Society of Ireland.

Dúchas, The Heritage Service (1999). *List of proposed candidate Special Areas of Conservation, Site Synopses, Habitat and Species lists.* Dublin: NPWS.

Dunford, B. (2001). *The Impact of Agricultural Practices on the Natural Heritage of the Burren Uplands, Co. Clare.* PhD Thesis, Unpublished. University College Dublin.

Dutton, H. (1808). *Statistical Survey of the County of Clare.* Dublin: Royal Agricultural Society.

Bibliography

Feighan, H. (1985). *Man's Impact on the Vegetation History of the Burren, Co. Clare.* MA Thesis, Unpublished. University College Cork.

Finch, T. F. (1971). *Soils of County Clare. Soil Survey Bulletin No. 23.* Dublin: An Foras Taluntais (The Agricultural Institute).

Foot, F. J. (1863). Explanation to accompany Sheets 114, 122 and 123 of the maps of the Geological Survey of Ireland. *Memoirs of the Geological Survey of Ireland.*

Grayson, F. W. (2000). The financial and ecological implications of restoring grazing regimes to grassland of high nature conservation value suffering from long-term agricultural abandonment; a case study. In: Crook and Penning (Eds.) *British Grassland Society, Occasional Symposium No. 3.*

Hencken, H. O' N. (1938). Cahercommaun, a stone fort in County Clare. *Journal of the Royal Society of Antiquarians of Ireland 68, 1-82. Dublin.*

Hopkins, J. J. (1996). *British Meadows and Pastures.* Origin Unknown. Photocopy Collection PC14333 University College Dublin Library.

Ivimey-Cook, R. B. and M. C. F Proctor, (1966). Plant Communities of the Burren, Co. Clare. *Proceedings of The Royal Irish Academy* 64B, 211-301. Dublin: Royal Irish Academy.

Jones, D. S. (1983). The Cleavage between Graziers and Peasants in the Land Struggle, 1890-1910. In: Clark, S. and J. S. Donnelly (Eds.) *Irish Peasants Violence and Political Unrest,* 1780-1914, 374-418. Manchester University Press.

Jones, D. S. (1995). Graziers, Land Reform and Political Conflict in Ireland. The Catholic University of American Press, Washington D.C.

Jones, C. (1997). Final Neolithic/Early Bronze Age Occupation of the Burren, A Brief Review of the Evidence. *The Other Clare* 21, 36-38. Ennis: Shannon Archaeological and Historical Society.

Keane, S. (1990). *Vegetation and land use studies in the Eastern area of the Burren, Co. Clare.* Unpublished MSc Thesis, University College Galway, Department of Botany.

Kent, M. and P. Coker (1992). *Vegetation Description and Analysis- A Practical Approach.* London: Belhaven Press.

Lewis, S. (1837). *County Clare: A History and Topography.* Republished 1995. Ennis: Clasp Press.

Lynch, A. (1988). Poulnabrone- A Stone in Time. *Archaeology Ireland* 2 (3), 105-107. Bray.

MacMahon, M. (1997). On a Fertile Rock: The Cistercian Abbey of Corcomroe. *The Other Clare* 21, 22-31. Ennis: Shannon Archaeological and Historical Society.

Bibliography

McCormick, F. (1995). Cows, Ringforts and the Origins of Early Christian Ireland. *Emania- Bulletin of the Navan Research group* 13, 33-38. Belfast: Queens University.

Mytum, H. (1982). The location of early churches in northern County Clare. *British Archaeological Reports (British Series)* 102, 351-361. Oxford.

Nelson, E. C. (2000). *An annotated topographical checklist of the flowering plants, conifers, ferns and fern allies of The Burren Region*. Dr. E.C. Nelson and The Burren Tourism and Environment Initiative, 2000.

Ní Ghabhláin, S. (1995). Church and Community in Medieval Ireland: The Diocese of Kilfenora. *Journal of the Royal Society of Antiquarians of Ireland* 125, 61-84. Dublin.

Ó Dálaigh, B. (ed.) (1998). *The Strangers Gaze: Travels in County Clare* 1534-1950. Ennis: Clasp Press.

O'Donovan, J. (editing and translating) (1851). *Annals of the Kingdom of Ireland by the Four Masters*. Dublin. Hodges, Smith and co.

Ó Drisceoil, D. (1988). Burnt Mounds: cooking or bathing? *Antiquity* 62, 671-680. Antiquity Publications Ltd. Oxford University Press.

O'Grady, S. H (ed.) (1929) *Caithreim Thoirdhealbhaigh: The Triumphs of Turlough*. J. MacRory-MacGrath. Irish Texts Society XXVII. London: Simpkin, Marshall Ltd.

Pienkowski, M. W. (ed.) (1998). Forum: Biodiversity and High-Nature-Value Farming. Conservation of Biodiversity by Supporting High-Nature-Value Farming Systems. *Journal of Applied Ecology*, 35, 948-973 British Ecological Society.

Robinson, T. (1999). *Introduction to the 1999 edition of the Burren map*. Folding Landscapes.

Rodwell, J. S. (ed.) (1992). *British Plant Communities Vol.3: Grasslands and Montane Communities*. Cambridge University Press.

Simington, R.C. (1641). *Book of Survey and Distribution*. Dublin: Irish Manuscripts Commission/ The Stationary Office.

Simms, M. (2001). *Exploring the Limestone Landscapes of the Burren and the Gort Lowlands*. Eden House, Belfast: Burrenkarst.com.

Traffic International (1999). *On Stony Ground: An investigation into trade in water worn limestone between the United Kingdom and the Republic of Ireland*. A report for the Countryside Agency and The Heritage Council. Cambridge: Traffic International.

Webb, D. A. and M. J. P. Scannell (1983). *Flora of Connemara and The Burren*. Royal Dublin Society and Cambridge University Press.

Other useful references

Comber, M. (ed.) (1999). *Archaeology of the Burren, T.J. Westropp. Prehistoric Forts and Dolmens in North Clare*. Ennis: Clasp Press.

Dunford, B. and J. Feehan (2001). Agricultural practices and natural heritage: a case study of the Burren uplands, Co. Clare. *Tearmann 1*(1), 19-34. Dublin: Teagasc and the Faculty of Agriculture, University College Dublin.

Nelson, E. C. and W. Walsh (1991). *The Burren: a companion to the wildflowers of an Irish limestone wilderness*. Kilkenny: Boethius Press and The Conservancy of the Burren.

O'Connell, J. W. and A. Korff (2002). *The Book of the Burren. 2nd Edition*. Kinvara: Tir Eolas

O'Donovan, G. (1987). *An Ecosystem Study of Grasslands in the Burren National Park, Co. Clare*. Unpublished PhD thesis, University College Dublin.

O'Donovan, G. (2001). Productivity, grazing pressure and phenology of a Sesleria-dominated grassland in the Burren National Park. *Tearmann* 1(1), 35-54. Dublin: Teagasc and the Faculty of Agriculture, University College Dublin.

Glossary

AAU: Agricultural Area Used.
AFF: Agriculture, Forestry and Fishing.
Bullock: A castrated male bovine.
Carboniferous: Geological period extending from 360 to 290 million years ago.
DED: District Electoral Division.
Doline: a closed natural depression of varying dimension.
Dry cow: A cow not producing milk.
Erratics: boulders picked up and rounded by glaciers, then dropped elsewhere.
Ewe: An adult female sheep.
Fencing: the stockproofing of field boundaries.
Ha: Hectare, equals 2.47 acres.
Heifer: A cow more than one year old that has not yet calved.
Heritage: what is, or may be, inherited, natural, man-made or cultural.
IFA: Irish Farmers Association, the main farming union in Ireland.
Karren: a range of small-scale solutional features on rock.
Karst: landscape features formed by the weathering of soluble rocks.
LU: Livestock Unit, equivalent to one cow (e.g. one sheep equals 0.15LU).
NHA: Natural Heritage Area.
REPS: Rural Environmental Protection Scheme.
RD: Rural District, political entity that encompasses a number of DEDs.
SAC: Special Area of Conservation, part of the Natura 2000 European network.
Slatted houses: livestock housing with effluent (slurry) collection facilities.
Store cattle: Cattle kept on a low level of growth (often over winter) prior to fattening or 'finishing' when grass/fodder becomes more readily available.
Suckler Cow: A cow the primary function of which is to produce and nurture offspring.
Teagasc: The Irish Agriculture and Food Development Authority.
Transhumance: from *trans humus*, the seasonal movement of livestock to another area.
Turlough: a seasonal lake of limestone regions that fills in, and drains out, from the base.
TWINSPAN: Two Way Indicator Species Analysis. A software programme used for classifying datasets (such as vegetation data).
Weanling: Young calf (under one year old).
Winterage: Grasslands traditionally used for grazing mainly from October to April (inclusive); also the practice or tradition of outwintering animals on grassland.